– Timothy Green Beckley's –

ALIEN ARTIFACTS

Incredible Evidence of Exotic Material From UFO Encounters

Dedicated to Our Friend Timothy Green Beckley (Mr. UFO).

God Speed With Your Journey on the Mothership.

– Timothy Green Beckley's –

ALIEN ARTIFACTS

Incredible Evidence of Exotic Material From UFO Encounters

ZONTAR PRESS

Alien Artifacts

Incredible Evidence of Exotic Material From UFO Encounters

By Sean Casteel & Tim R. Swartz

Contributors: Scott Corrales, Tom Hackney, Hercules Invictus, Mark Olly, Calvin Parker, Paul Dale Roberts, Alejandro Rojas, Gene Steinberg, Lon Strickler, Diane Tessman, Nigel Watson

First Edition
First Printing, 2022

Published in the United States of America
By Zontar Press

www.conspiracyjournal.com

CONTENTS

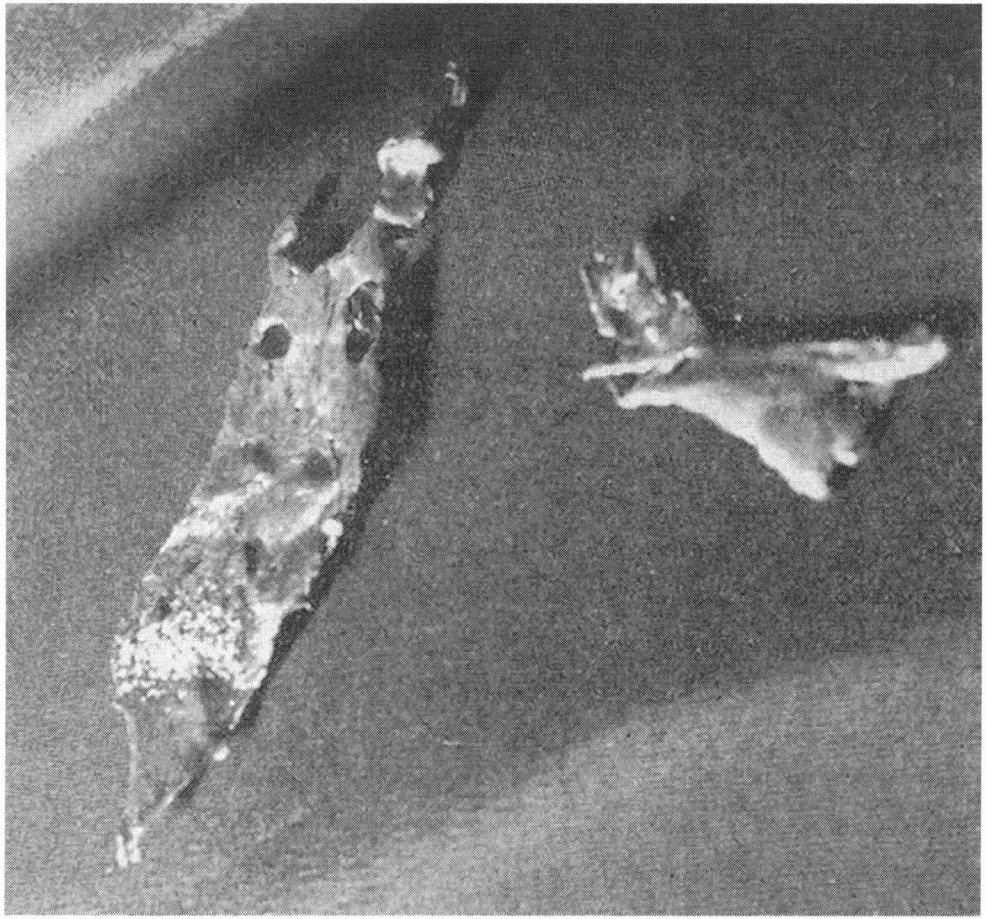

On December 14, 1954, three disk-shaped objects were seen flying over the city of Campinas, Brazil. Suddenly, one of the disks started wobbling and lost altitude. The other two disks followed the wobbling UFO as it stabilized at an altitude of 300 ft. At this point the unstable disk emitted a thin stream of silvery liquid which was reported to splatter over roofs, streets, sidewalks, and even clothes left outside to dry. An analysis of the samples by an unnamed Brazilian government lab identified tin as the main component of the collected samples. This was also confirmed by a private chemist, Dr. Risvaldo Maffei, who also reported that around 10% of the samples contained other substances along with the tin, but what those other substances were, was not given.

Above: George Pedley, along with Albert Pennisi, examines reeds found floating within a circle of clockwise swirled reeds, measuring 30 feet in diameter, on the surface of a lagoon. Pedley had reported seeing a "flying saucer" rise from the lagoon the night before.

Below: One of the "saucer nests" of Tully, Australia, January 19, 1966.

INTRODUCTION: WHAT THE STRANGERS LEAVE BEHIND

By Tim R. Swartz

A Congressional Hearing was held on May 17, 2022, by the House Subcommittee on Government Investigations of Unidentified Flying Objects (UFOs). It was the first meeting of its kind to take place in more than 50 years. Afterwards, Congressman Tim Burchett called the UFO hearing a "total joke."

"We should have heard from people who could talk about things they'd personally seen, but instead the witnesses were government officials with limited knowledge who couldn't give real answers to serious questions," Burchett said in a Tweet.

Prior to the May 17 hearing, Burchett also claimed that wreckage from UFOs had been recovered. He told reporters that "multiple sources" had informed him of the fact but did not elaborate further.

Recovered UFO wreckage? Where have we heard that one before?

It is rare that we would have a U.S. Congressman making such an inflammatory statement. But, like past proclamations about recovered UFOs (or even just some mysterious debris), we are given a lot of hyperbole, but do we have any actual, physical evidence from UFOs...or what we think were UFOs?

That is what this book is all about. We have named it "*Alien Artifacts,*" but actually we are using this term not so much to say that we believe that we have recovered things from "extraterrestrial spaceships," but because what we are discussing in this book is "Alien" as they don't appear to have a "normal" explanation of their origins.

On top of that, we are also going beyond the strange bits of metal that have surfaced following UFO encounters. An "Alien Artifact" can be a lot of different things and we want to open your eyes to that fact.

What you won't find too much in this book are cases of crashed saucers and Roswell. Oh, they'll be discussed from time to time, but we feel that this subject has been done to death in dozens of other books and articles and we want to get past all of that into some more rarely explored territory.

Here is a good example of "Alien Artifacts" that aren't chunks of metal that look like they came off a mangled soda can. This is the case of Joe Simonton, the Wisconsin farmer who in 1961 was shocked to see a metallic "flying saucer" land outside his house. To him, it looked like two inverted bowls with "exhaust pipes" at the ends.

The farmer rushed outside and saw a hatch opening from the craft, "just like the trunk of your car."

Inside, Simonton saw "little men" about five feet tall, "dark-haired, dark-eyed, dark skin." In dress, they wore black or navy blue turtleneck shirts and helmets. However, the "chef" had red-striped pants.

According to reports, the three beings looked to be 20 to 30-year-old and "Italian" in appearance.

Gesturing to Simonton, the man held up a metallic jug, indicating that he would like to have it filled with water. Simonton took the jug, filled it with water and returned it to the little man.

FOOD OF THE GODS

Simonton looked inside the saucer and saw another little man cooking "pancakes" on a smooth, square grill-like surface. Hoping to "get a conversation out of him," the farmer gestured that he would like to try the food. In return, the pilot handed Simonton four of the pancakes, which were "hot and greasy."

Being polite, Simonton tried the pancake.

"If that was their food, God help 'em," said Simonton. "Because I took a bite of one of 'em and it tasted like a piece of cardboard. And, if that's what they lived on, no wonder they're small."

The pilot then indicated that Simonton should step back and closed the hatch. The craft rose up vertically, "like an elevator," Simonton remembered, and it shot off into the sky and quickly disappeared.

"Everything was timed perfectly," Simonton said. "It went up about 20 feet. It tilted at a 45-degree, straight south, and took off. Within two to three seconds it was out of sight."

Joe Simonton shows off one of his "alien pancakes."

It's an amazing story...but unlike the thousands of other UFO reports, this one came with actual, physical evidence in the form of strange, bubbly textured pancakes.

After the story got out, the Air Force sent J. Allen Hynek to look into the case. According to Hynek, the man's story was genuine.

"There is no doubt that Mr. Simonton felt that his contact was a real experience," said Hynek.

Later, a statement from the Air Force suggested they had the sample tested and they were ordinary "buckwheat pancakes" consisting of fat, starch, buckwheat hulls, wheat bran and soybean hulls. Officially, the Air Force labeled the case as "unknown."

Like many of the cases we will explore in this book, Simonton did seem to have an encounter of some type with the unknown, but the physical evidence left behind did not appear to be of extraterrestrial origin...that is unless alien races are also growing buckwheat on some distant planet.

SCARRED BY A UFO

Another example of an "alien artifact" that goes beyond the traditional are cases where a close encounter leaves the witness with a physical injury. This brings to mind the Falcon Lake incident of May, 1967.

Stefan Michalak was an industrial mechanic by trade and an amateur geologist who liked to venture into the wilderness around Falcon Lake — about 93 miles east of Winnipeg — to prospect for quartz and silver.

ALIEN ARTIFACTS

On May 20, 1967, Michalak was chipping away at a vein of quartz when he looked up and saw disc-shaped objects with a reddish glow hovering in the sky.

One object descended and landed on a nearby flat section of rock. The other remained in the air for a few minutes before flying off.

Believing it to be a secret U.S. military experimental craft, Michalak sat back and sketched it over the next half hour. When a hatch opened on the side, he decided to get a closer look. At the time he was wearing welding goggles to protect his eyes when he was prospecting, and it was a good thing he did, because peering inside, Michalak recalled that there were blindingly bright lights everywhere, focused beams shining across the interior along with a series of flashing lights turning on and off in seemingly random sequences.

After hearing muffled voices from the craft, he called out, offering mechanical help to the "Yankee boys" if they needed it. The voices went quiet, so Stefan tried in his native Polish, then in Russian and finally in German.

He reached out to touch the smooth surface of the craft, which he said melted the fingertips of the glove he was wearing.

The craft then began to turn counter-clockwise, which brought up a panel that contained a series of holes spaced in a grid pattern directly in front of him. Suddenly, Michalak was struck by a scalding hot blast of gas that threw him backwards and set his shirt and cap on fire.

Ripping away the burning garments, Michalak saw the craft lift off and fly away. Disoriented and nauseous, Stefan stumbled through the forest and vomited. He eventually made his way back to his motel room in Falcon Lake then caught a bus back to Winnipeg.

Police officers went to the location of the incident and found Michalak's burnt shirt, cap and other personal belongings. An investigation conducted in 1968 found traces of radioactivity in a semicircle around the center of the landing site. Also found were pieces of metal that had to be chipped out of cracks in the rock. The metal had somehow been melted into the cracks. Tests showed the presence of radium-226, an isotope of radium that is used for various commercial purposes and is also be found in waste from nuclear energy.

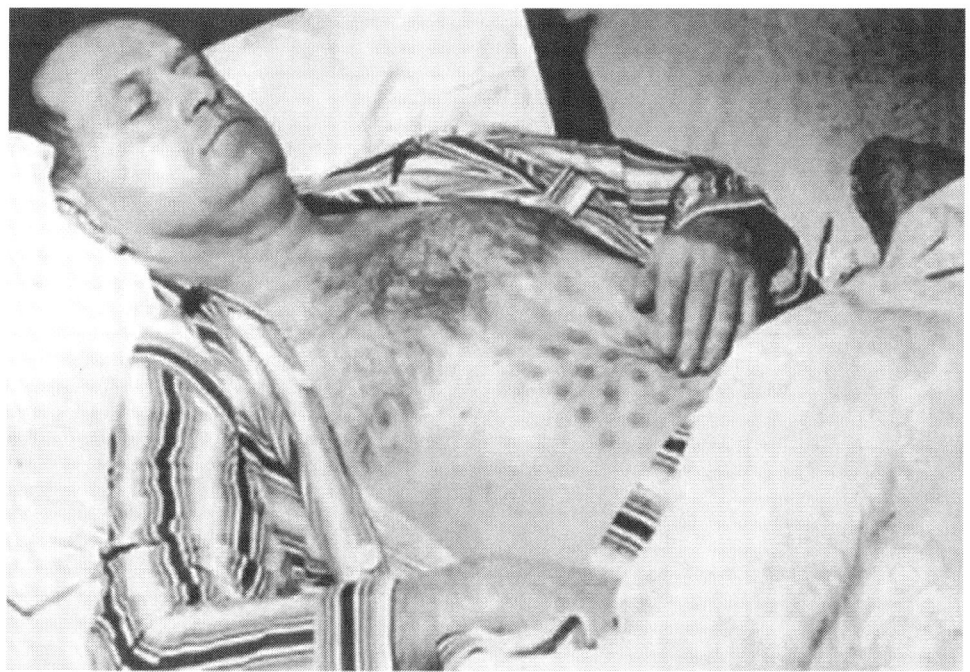

Stefan Michalak suffered mysterious burns to his abdomen after a close encounter with a UFO.

Michalak was treated at a hospital for burns to his chest and stomach that later turned into raised sores on a grid-like pattern. And for weeks afterwards, he suffered from diarrhea, headaches, blackouts and weight loss. In fact, he suffered the remainder of his life from his frightening encounter. According to his son, Stan Michalak, the strange burns on his father's chest would reappear several times a year accompanied by other debilitating symptoms. This prompted him to check himself into the Mayo Clinic in 1968.

Doctors did a thorough investigation and even sent him to a psychiatrist "who came back with the report that this is a fellow who's very pragmatic, very down to earth." Unfortunately, other than a vague suggestion that Michalak had been exposed to some kind of radiation, nothing could be found to explain his baffling symptoms.

ALIEN TOURISTS

Whenever tourists visit a scenic location, there are always a few inconsiderate ones that leave something behind, usually trash and garbage. But according to an account by Brent Swancer on the *"Mysterious Universe"* website, a group of possibly out-of-this-world tourists left behind some interesting evidence of their holiday on Earth.

Sometime in 1970, a hotel employee in St. Louis, Missouri, told investigator John E. Schroeder about a group of odd "tiny beings" at their establishment that they believed to be not of this world.

Dorothy Simpson had been sitting at the reception desk of the hotel when she looked up at a group of very strange-looking individuals standing in front of her.

Simpson said they very small, barely at eye level with the edge of the desk, with pale, slightly triangular faces that started wide at the eyes but drew thinner down to pointy chins, which held tiny, lipless mouths. The eyes themselves were large, dark and slightly slanted, and their noses were nearly nonexistent, little more than two slits. She also said that they all appeared to be wearing very bad looking wigs.

Although they were rather androgynous in physical appearance, two were in expensive tailored men's suits and the others were dressed in pastel peach dresses, but if not for the clothes and the different lengths of their hair there would have been no way to tell who was male and who was female. Two of them seemed slightly smaller than the others, giving the impression that they were perhaps children, but it was hard to tell.

One of the "men" spoke out in a high-pitched voice to ask for a room. But when Dorothy told him the price he didn't seem to understand. He needed help from one of the women to realize that it meant he needed money. He then pulled out a crisp stack of bills from his pocket and paid the required amount.

When she asked the gentleman's name, he told her he was "A. Bell," and she helped signed the register for him as he was too short to reach out over the desk to do it himself. When asked where he was from, he pointed up at the sky, but the woman next to him gently lowered his arm and told her they were from Hammond, Indiana, even providing an address.

Afterwards, other employees had noticed just how weird the visitors were, which prompted the motel to check on the Indiana address, which turned out to be fake. He also did a test of the extremely new bills they had been given on suspicion that they were counterfeit, but those proved to be real. They also checked the parking area for any car with Indiana plates but there was none.

Later that evening, the bellhop found their strange visitors wandering around in a confused state. Helping them find their room, the bellhop was scolded by one of the women for turning on the lights too bright. The next day, the visitors had vanished, despite no one seeing them leave. It was as if they had just evaporated into thin air.

The only thing left behind as proof of their visit was the crisp, new money they used to pay for their rooms, and a lot of unsettled employees wondering what to make of their bizarre experience.

• • •

You can see that "alien artifacts" can show up in all sorts of shapes and sizes, from exotic material to just plain old mundane things that we encounter in everyday life, but strangely appearing in not so mundane situations.

For this book we have gathered some of our favorite authors and researchers to contribute chapters on a subject that offers no easy explanations. Alien artifacts, their origins and meanings, continue to mystify us and probably will for years to come.

ALIEN ARTIFACTS

Experts Think A Strange Metal Object Could Be Part of an Ancient UFO

In 1973, builders working on the shores of the Mures River not far from the central Romanian town of Aiud found three objects 33 feet underground.

They appeared to be unusual and very old, and archaeologists were bought in who immediately identified two of them as being fossils. The third looked like a piece of man-made metal, although very light, and it was suspected that it might be the end of an axe.

All three were sent together with the others for further analysis to Cluj, the main city of the Romanian region of Transylvania.

It was quickly determined that the two large bones belonged to a large extinct mammal that died 10,000-80,000 years ago, but experts were stunned to find out that the third object was a piece of very lightweight metal, and appeared to have been manufactured.

Archeologists enlisted to study the aluminum initially thought it might be the end of an ace, but further study found it was an extremely lightweight metal that had been manufactured.

Tests showed that the object is made of 12 metals, 90% aluminium, and it was dated by Romanian officials as being 250,000 years old.

More experts brought in to conduct tests said the dates were far later, ranging between 400 and 80,000 years old - but even at 400 years old it would still be 200 years earlier than when aluminum was first produced.

ALIEN ARTIFACTS

The object has concavities that suggest that it was part of a more complex mechanical system - and possibly alien in origin.

Gheorghe Cohal, the Deputy Director of the Romanian Ufologists Association, told local media: "Lab tests concluded it is an old UFO fragment given that the substances it comprises cannot be combined with technology available on Earth."

Local historian Mihai Wittenberger has tried to claim that the object is actually a metal piece from a World War II German aircraft - but this does not explain the age of the artifact.

The metal object has now gone on display in the History Museum of Cluj-Napoca with a sign alongside that reads "origin still unknown."

Source: Yahoo News - https://ca.news.yahoo.com/experts-think-ancient-metal-object-142234156.html

The mysterious aluminum wedge of Aiud.

1.

THE ALIEN GLITTERING OF TRUE GOLD
By Sean Casteel

Some ancient peoples seem to have spared no expense when it came to creating their works of art, using the medium of gold in their pursuit of a higher form of spiritual truth. For example, there are the Quimbaya of Columbia, South America, who were believed to inhabit the region from 300 to 1550 CE and are best known for their precise gold and metal work.

According to a website called *"Mystery Pile,"* the majority of the Quimbaya pieces discovered by archeologists were an alloy of gold and 30 percent copper, called "tumbaga," and very similar to a metal mentioned by Plato in his dialogues about the Lost City of Atlantis. The objects included golden models of several types of insects along with devices that are aerodynamic in nature and shaped like no insect known to exist.

"The ancient pieces look very much like the designs of modern airplanes," the site says, "and incorporate a number of features essentially proving the Quimbaya knew and understood principles of flight."

OUT OF PLACE IN TIME

Another website, called *"The Gypsy Thread,"* says that "The overtly stylized gold objects measure between two and three inches, with each piece customized to a unique look. Researchers have classified them as depictions of lizards, butterflies, birds and insects common to the area, yet it's unmistakable that they also look like many of our modern-day flying machines – some complete with tail rudders and propellers. The existence of so many similarities to modern-day airplanes supports the 'Out of Place Artifact' theory: they seem too advanced for the Quimbaya."

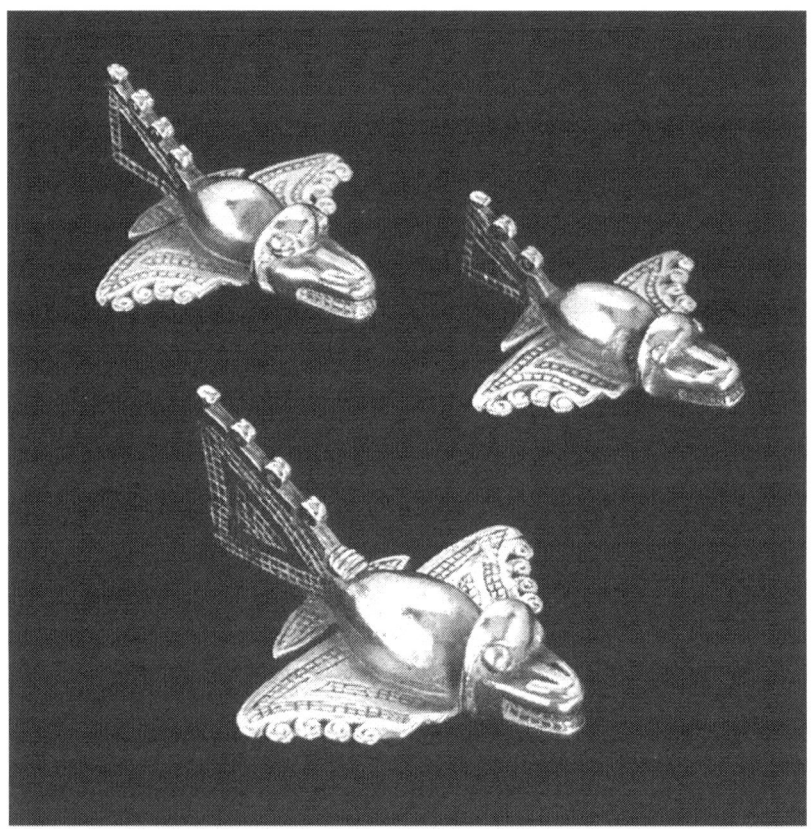

Some of the Quimbaya objects look very much like the designs of modern airplanes and incorporate a number of features that could show the Quimbaya knew and understood principles of flight.

The Out of Place Artifacts theory is one that challenges the historical record in some way, shape or form, "The Gypsy Thread" goes on. These can be items which seem too advanced for a particular civilization or, in some cases, items which show a human presence when no humans were supposed to exist.

"The term is rarely used by scientists or archeologists" the site continues, "but widely accepted by those who believe in ancient astronaut scenarios, students of the paranormal and UFO enthusiasts."

While the scientific community has successfully refuted many claims and shown many items to be hoaxes, there are some artifacts that are still impossible to connect to the time period they are associated with.

One of the problems with properly classifying the Quimbaya objects stems from the fact that they weren't discovered through normal archeological processes. They were looted in the late 1800s from an area known as the Central Cauca Valley. While some archeologists believe the items came from two tombs, they cannot say so with 100 percent certainty. The current collection of 123 items only exists because someone turned them in to the Columbian authorities. It's almost certain that many similar items from the region are held in private collections throughout the world.

WHO WERE THE QUIMBAYA?

The Gypsy Thread" website offers an historical account of the civilization that created the anachronistic objects.

"The Quimbaya civilization inhabited the area around the Cauca River Valley on the western slopes of the Andes Mountains. There is no clear evidence that pinpoints when the Quimbaya came into being. However, most researchers agree it was sometime in the first century B.C. They were expert hunters, grew many different and diverse crops, fished and had many industries, including gold mining and gold-smithing. The Quimbaya civilization reached its peak in the period between the fourth and seventh century A.D. Spanish Conquistadors began to colonize Columbia in 1509, which led to the end of the Quimbaya period. The people were known for their spectacular gold work with highly detailed and unique designs."

ACTUAL FLIGHT HAS BEEN ACHIEVED

It has been theorized for years that some of the Quimbaya artifacts are scale models of airplanes or flying machines. Two aeronautical engineers, Peter Belting and Conrad Lubbers, used the dimensions of the Quimbaya artifacts to create large-scale models which proved successful in flight testing. They proved that the designs fly with both simple single-propeller power and jet power.

PETROGLYPHS IN TRIBUTE?

Another layer of mystery is added by petroglyphs and stone carvings in hard granite in an area where the Quimbaya lived.

"In places like the Park of the Marked Stones," "*The Gypsy Thread*" says, "and Natural Park of Las Piedras Marcadas, carvings seem to support some knowledge of constellations and the stars. Little else is known about these carvings, including the date they were made or their true meaning. Some theorize they were made in honor of extraterrestrial encounters."

In any case, the artifacts are clearly the work of master craftsmen with an eye for detail, but again seem out of place for the time period. It's more than just a coincidence that the artifacts can be scaled up and actually fly.

"The Mystery Pile" website adds another interesting idea to the mix.

"Today we find a distinctly intriguing phenomenon," the site says, "which takes place after a remote culture is visited for the first time with modern technology present. Isolated tribes visited in both Africa and South America by airplanes have both demonstrated shifts in religious beliefs after the visit. One of the tribes welcomed the plane on its second visit with ceremonial fire and statues constructed in the shape of the airplane. Tribal people even went so far as to line themselves along a runway to greet the visitors. If remote cultures exhibit this sort of behavior during the world's modern technological era, then likely the same concept has played out before. From this angle of thinking, theories then suggest that the Quimbaya may have been influenced by another ancient culture or, perhaps, some sort of alien civilization."

FROM THE BRONZE AGE COME HATS OF GOLD

Another example of what may prove to be an alien fondness for gold comes in the form of golden hats believed to endow the wearer with divine powers. The story is told by the website "Ancient Origins."

"Occasionally, an astonishing find challenges our understanding of ancient societies and cultures and provides surprising new information about civilizations of the past," the site says. "One such find was the discovery of four cone-shaped golden hats from the Bronze Age. Discovered in different locations and at different times, the four gold hats share many similarities in size, shape, design and construction.

"Their conical design," the site continues, "mimics the well-known image of a witch's or wizard's hat, leading to speculation that the hats were worn by individuals who held such a position. The hats are engraved with symbols that may have been used to make agricultural and/or astronomical predictions, possibly raising the wearer to divine status."

The four hats date back to the Bronze Age, which lasted from 3300 to 700 BC, and they all appear to have been created sometime around the middle of that period, ranging from 1400 to 800 BC. They were each discovered separately, over the course of 160 years. Three were found in Germany and the fourth one in France. It is also possible that more gold hats will be uncovered at some future point.

BRIEF DESCRIPTIONS OF THE FOUR HATS
The golden relics are constructed of sheets of gold, with intricate astronomical designs, and demonstrate superb craftsmanship. While the four hats are strikingly similar in some ways, they are also somewhat unique in their specific features.

The First Hat: Discovered in 1835 at Schifferstadt, Germany. It is called the Golden Hat of Schifferstadt. Uncovered by a farmer, it appeared to have been intentionally buried. It is the shortest of the four hats at 29.6 cm high. It is divided into bands that run the full length of the hat, with each band decorated with one of several designs, including circles, disc shapes and eye-like shapes. It is believed to have been manufactured sometime between 1400 and 1300 BC.

The Second Hat: Discovered in Avanton, France, in 1844, it is called the Avanton Gold Cone. It is believed to date from 1000 to 900 BC, and is the only hat missing a brim, though damaged remains indicates that it once did have a brim. It stands at 55 cm. and is also banded with repeated circle symbols.

The Third Hat: Called the Golden Cone of Ezelsdorf-Buch, named after the German town where it was found in 1953, it stands as the tallest of the four hats at 88 cm. containing the same banded design as the other hats, with repeated circles, discs and eye-like shapes, it is believed to have originated between 1000 to 900 BC.

The Fourth Hat: The exact place of origin is less clear with this hat, but it is believed to have been found in either southern Germany or Switzerland. It is dated to approximately 1000 to 800 BC, and is known as the Berlin Gold Hat because it was purchased by the Berlin Museum. It features the same banded pattern as the others and stands at 75 cm tall.

WHAT ARE THE HATS FOR?

According to the *"Ancient Origins"* website, the purpose of the hats is unknown. While the hats were each found in different regions, the hats are thought of as a group under the assumption that they were all used for similar purposes.

The Four Golden Hats are made of thin sheet gold and were attached externally to long conical and brimmed headdresses which were probably made of some organic material and served to stabilize the external gold leaf.

The phallic shape of the hats led to the belief that they may have been intended to be symbols of fertility. Some researchers thought the hats may have been part of a suit of armor or were used as ceremonial vases. It has also been speculated that the four hats may have belonged to ancient wizards, due to their resemblance to wizard-style hats.

More recent theories advance the idea that the astrological symbols on the hats were used to track the stars and the sun, which allowed for agricultural predictions, like when to plant and harvest. The figures who wore the hats were referred to as "king-priests" because they were able to make predictions and were therefore believed to have supernatural powers. While predictions of time and weather are commonplace today, due to modern knowledge and technology, the ability to predict climate conditions during the Bronze Age was seen as a divine power.

Wilfried Menghin, the director of the Berlin Museum, theorized that the king-priests "would have been regarded as Lords of Time who had access to a divine knowledge that enabled them to look into the future."

The use of the four gold hats to predict the movements of the sun and the time relationship between the sun and the moon isn't entirely new, as many other ancient artifacts also focus on such astrological elements. But why they chose to express such knowledge on golden hats is still a mystery. It remains to be seen why the "wizards" of the Bronze Age wore these remarkable hats of gold.

The mysterious stone spheres were first discovered in the 1930s when workmen were clearing their way through the dense jungle of Costa Rica for banana plantations.

2.

LIKE A ROLLING SPHERE
By Sean Casteel

The ancient stone spheres of Costa Rica were made world-famous by the opening sequence of *"Raiders of the Lost Ark,"* when a mockup of one of the mysterious relics nearly crushed Indiana Jones, says a website called *"Science Daily."*

The article includes an interview with University of Kansas anthropology professor John Hoopes, who researches ancient cultures of Central and South America and is one of the world's foremost experts on the Costa Rican spheres. He explained that, although the spheres are very old, international interest in them is still growing.

The *"Science Daily"* article quotes Hoopes as saying, "The earliest reports of the stones come from the late 19th century, but they weren't really reported scientifically until the 1930s – so they're a relatively recent discovery. They remained unknown until the United Fruit Company began clearing land for banana plantations in southern Costa Rica."

According to Hoopes, around 300 balls are known to exist, with the largest weighing 16 tons and measuring eight feet in diameter. Many of these are clustered in Costa Rica's Diquis Delta region. Some remain pristine in the original places of discovery, but many others have been relocated or damaged due to erosion, fires and vandalism.

Hoopes said that scientists believe the stones were first created around 600 AD, with most dating to after 1,000 AD, but before the Spanish conquest.

"We date the spheres by pottery styles and radiocarbon dates associated with archeological deposits found with the stone spheres," Hoopes said. "One of the problems with this methodology is that it tells you the latest use of the sphere, but it doesn't tell you when it was made. These objects can be used for centuries and are still sitting where they are after a thousand years. So it's very difficult to say exactly when they were made."

A report on the *"Crystalinks"* website says that the stone spheres came to light during the early cultivation of the farmland. Most were discovered by workmen as they cleared and burned the jungle in preparation for planting. Recognizing the stones as manmade, the workmen pushed them aside with bulldozers and heavy equipment.

At some point thereafter, they returned to the stone balls and, inspired by stories of hidden gold, began to drill holes into them. They inserted sticks of dynamite – which they would normally use to remove stubborn roots and stumps – into the holes, destroying several of the spheres before authorities intervened. Some of the dynamited spheres were reassembled and exhibited at the National Museum in San Jose.

One of the stranger beliefs about the spheres is that they are a remnant of the Lost Continent of Atlantis. Another is that extraterrestrials guided the process by which the spheres were made for some unknown purpose.

The larger stones were clearly crafted by the most skilled sculptors. These stones are so perfectly shaped that the tape and plum bob measurements of diameters reveal no imperfections. This shows that the makers of the stones must have had a degree of mathematical ability as well as advanced knowledge of stone carving and the use of tools.

"The ancient Costa Ricans, however, had no written language," the "*Crystalinks*" site says. "There is, therefore, no written record of just how they made the stone spheres."

Have Researchers Found 300,000-year-old Nano-structures In The Ural Mountains?

Found in the Ural Mountains in Russia, these objects have caused quite a buzz since their discovery. The tiny structures are believed to have been the product of an extremely ancient civilization that was capable of developing nanotechnology about 300,000 years ago.

The objects were discovered during a geological research mission whose purpose was the extraction of gold in the Ural Mountains in Russia, and while gold was their main focus, researchers were amazed to find something apparently much more valuable. The pieces discovered are coils, spirals and shafts among the list of unidentified components that were unearthed during the geological missions in the area.

The Russian Academy of sciences performed several tests on these mysterious objects and the results were quite interesting. Researchers found out that the largest pieces that were unearthed were made almost entirely out of copper and the smaller ones from tungsten and molybdenum.

The Russian Academy of Science has a structure of 11 specialized scientific divisions, three territorial divisions also referred to as branches, and it consists of 14 regional scientific centers. The Academy has numerous councils, committees and commissions, organized for different purposes and studies.

A lot of people who read about these artifacts have discredited their discovery and meaning stating that there is no research facility from the RAS and that the research performed was dubious but they are mistaken as the Ural Division of the RAS was established in 1932, with Aleksandr

Fersman as its founding chairman. Research centers are in Yekaterinburg, Perm, Cheliabinsk, Izhevsk, Orenburg, Ufa and Syktyvkar.

The materials were submitted to a more extensive research a couple of years after their discovery to find out more about the mysterious objects and their composition. According to the Russian Academy of Science and their department for Geology; the metals did not originated in nature on their own, meaning that they are components that have an artificial technological origin, in other words they were manufactured.

According to the information available, these nanostructures were found at a depth between 10 and 40 feet, research also shows that they range in date from 20,00 to over 300,000 years. But who made them and for what purpose? Are they the result of a now extinct ancient civilization? Or are they the result of something beyond our planet? These are the questions that have been asked when it comes to the origin of these ancient nanostructures.

Even though research has been made in Russia, some argue that given the skeptical opinion worldwide about these findings, it would have been interesting to see what other researchers in other countries would have to say about these nanostructures. Even though the materials were studied in Helsinki, St. Petersburg and Moscow, there have been no documents made available concerning the nanostructures, their origins and purpose.

According to Dr. E.W. Matveyeva, of the Section for Geology, Prospecting Techniques, and Economics of Precious

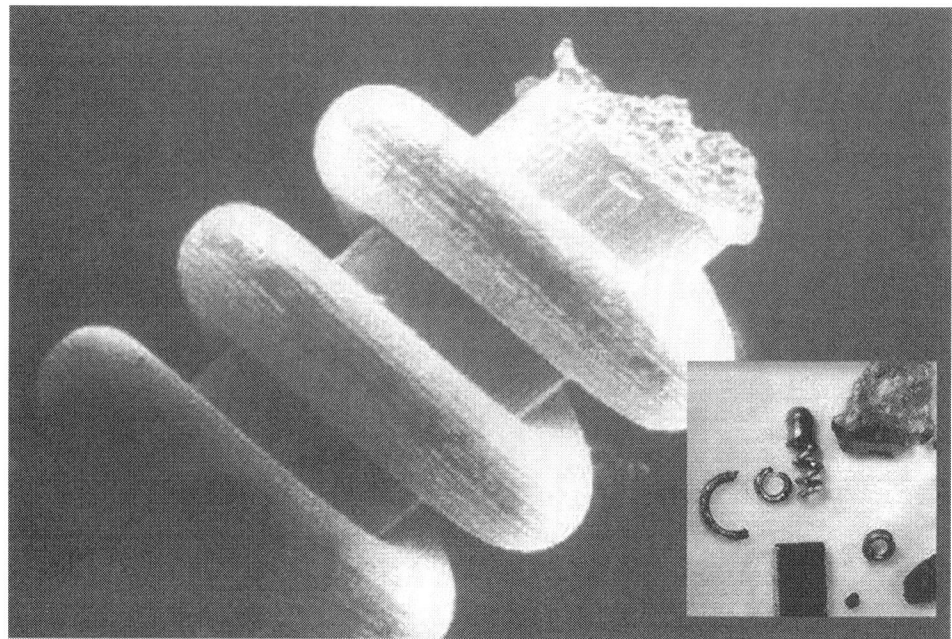

In 1991, Russian prospectors first discovered tiny metal coils buried along the banks of the Naroda river.

Metal Alluvial Deposits; the layer which contains the spiral-shaped objects is characterized as gravel and detritus deposits of No. 3 stratum, which in our view, show inner-sedimentary erosion of polygenetic accumulative layers (i.e. layers composed of material of various origins). From their orientation these layers can be dated to 100,000 years and correspond to the the lower regions of the Mikulinsk horizon of the upper Pleistocene.

Particular attention should be paid to the final conclusion reached by the Moscow institute. Report No. 18/485 states that the age of the deposits and the results of the tests give a very low probability to the assumption that the origin of these unusual, thread-shaped tungsten crystals is of a technogenic cosmic nature, due to the rocket launch route from the Plesetsk Space Center which goes over the northern part of the Ural region.

Debate over these objects will continue due to the "mysterious" nature of these objects and while there is a possibility that these nanostructures could have originated from an ancient civilization that lived on Earth hundreds of thousands of years ago, other possibilities cannot be excluded.

Source: Ancient Code

www.ancient-code.com/300000-year-old-nanostructures-found-in-the-ural-mountains/

The British humor magazine "Punch" satirizes the 1909 "Airship Panic" which saw reports of strange objects in the sky over Great Britain.

3.

PHANTOM AIRSHIP WRECKAGE
By Nigel Watson

Rumored activities of German secret agents were very much linked in the public mind with the British phantom airship sightings of 1909 and 1913. This kind of link, and other stories recorded during these periods, appears to be very similar to some of the more bizarre aspects of the contemporary UFO scene.

The ambiguity and bizarre nature of some of the airship sightings, the fear of alien invasion, the existence of foreign spies and mad inventors, allied with secret government investigations, in the 1900's, parallels the modern day UFO phenomenon, which also presents witnesses with strange encounters, the fear of alien (extraterrestrial) invasion, Men-In-Black, and secret government involvement.

The British airship wave of 1909 started getting press attention on 15 May, and fizzled out in early June. The same can be said of the accompanying spy scare.

A scenario which reveals a combination of airship and spy elements was played out in Clacton-on-Sea, Essex, when Mr. Egerton S. Free, in the early part of May, saw an airship hovering near his home. "I looked up," he said," and in the sky I saw a long, torpedo-shaped balloon, high up in the air overhead. It was a clear, fairly light night, and I could see everything most distinctly. The airship was travelling swiftly in the direction of Frinton and showing two bright lights. I stood and watched it for some time until it disappeared." (1)

The next day a 4 ft 6 inch India-rubber bag, which was presumed to have been dropped from the airship, was found.

Mr. Free said: "When I got to the cliff-edge at the top of the steps I saw a most curious shape sticking up in the sandy grass. It looked like a large, slightly flattened football, with a steel bar pushed right through it. I picked it up – it was not very heavy – and brought it into the house. The contrivance is about five feet long from end to end. The central bar is of hollow steel, with an end round and flat like the buffer of a railway engine. The ball part is about three feet long, oval-shaped, made of hard grey rubber, and corded net-fashion with twine, while the words *Muller Fabrik Bremen*' are painted on the ball in black letters. The steel ends project about a foot on either side of the ball, and when I found it the sharp end was sticking in the grass and the ball was on its side." (2)

The Navy took the bag away for examination. A few weeks later they identified it as a "reindeer buoy" that is used as a gunnery practice target.

On 16 May, the day after Mr. Free's sighting became publicly known; two strangers investigated the vicinity of his home. They looked at the area where he had seen the airship, and Mr. Free said: "The men hovered about my house persistently for five hours that is until 7 o'clock in the evening. When the servant girl set out for church she heard them conversing in a foreign tongue. Finally they came up to her, one on each side, and one of the men spoke to her in a strange language. The girl ... was so frightened that she ran back to my house, and would not again leave for church." (3)

According to UFO researcher David Clarke "The clear implication was that the 'foreigners' were German agents who had been dispatched to recover the 'buffer' dropped by the Zeppelin and so remove evidence of its spying mission. Cold water was later poured on these claims by a reporter from the *"East Essex Advertiser,"* a Clacton paper, who noted how a respected local amateur photographer had innocently gone to the house to take a picture of the object on the beach. It added: "The next day it was reported that a foreign looking gentleman with a camera had asked permission to take a photograph of the 'find.'" (4)

THE CAERPHILLY WRECKAGE

On 16 May, the same day as foreigners were seen by Mr. Free near Clacton, a stockbroker's clerk saw five foreigners on Caerphilly Mountain. They rode from spot to spot in two traps (carriages) photographing and surveying the area. They finished their work at midday, and one of the traps went on the road to Llanishen while the other one took the road to Cardiff. (5)

Their airship connection is with the sighting by Mr. Charles Lethbridge two days later on the same mountain. The question is left open as to whether they were surveying the area in preparation for an airship visitation or it was just a coincidence?

What we would now call a close encounter of the third kind occurred on 18 May, near the top of Caerphilly

Mountain, South Wales. Mr. Lethbridge, a 59-year-old Cardiff Punch and Judy showman, told reporters that: "I work during the winter months at the Cardiff Docks, but in the summer-time I travel the district with my little Punch and Judy show, giving performances at the various schools. Yesterday I went to Senghenydd, and after covering a few pitches, proceeded to walk home over Caerphilly Mountain.

"You know that the top of that mountain is a very lonely spot. I reached it at about 11 pm and when turning the bend at the summit I was surprised to see a long, tube-shaped affair lying on the grass on the roadside, with two men busily engaged with something nearby.

"They attracted my close attention because of their peculiar get-up. They appeared to have big, heavy fur-coats, and fur-caps fitting tightly over their heads. I was rather frightened, but I continued to go on until I was within twenty yards of them, and then my idea as to their clothing was confirmed.

"The noise of my little spring cart seemed to attract them, and when they saw me they jumped up and jabbered furiously to each other in a strange lingo – Welsh or something else, it was certainly not English.

"They hurriedly collected something from the ground, and then I was really frightened. The long thing on the ground rose up slowly – I was standing still at the time, quite amazed - and when it was hanging a few feet off the ground the men jumped into a kind of little carriage suspended from it, and gradually the whole affair and the men rose into the air in a zig-zag fashion.

"When they cleared the telegraph wires that pass over the mountain, two lights, like electric lamps, shone out and the thing went higher into the air and sailed away towards Cardiff. I was too frightened to move for a time, but I pulled myself together, and as soon as I came home told my people about what I had seen."

Lethbridge gave more details about the aeronauts:

"They were two tall, smart young men, and I am also certain that they did not speak English, for when they looked towards me they spoke very loudly to each other, as if quarrelling or excited, as I made up my mind at once that they were foreigners.

"When the thing went into the air I distinctly saw what looked like a couple of wheels on the bottom of a little carriage, and at the tail end of it was a fan whirring away as you hear a motorcar do sometimes." (6)

In support of his extraordinary encounter, workers at the Cardiff Docks said they saw an airship fly overhead during their supper break. A signalman at the King's Junction, Queen Alexandra Dock, Robert Westlake, stated: "At 1.15 this morning (19 May), while attending to my duties signaling trains...I was startled by a weird object flying in the air. In appearance it represented a boat of cigar-shape, and was making a whizzing noise. It was lit up by two lights, which could be plainly seen. It was travelling at a great rate, and was elevated at a distance of half-a-mile, making for the eastward. A number of men working on the steamship Arndale also saw the airship. It came from the direction of Newport, and took a curve over the docks, and passed over the Channel towards Weston, being clearly in view for a minute or two before the lights on board were suddenly extinguished." (7)

Members of the crew of the Arndale also saw the craft. A workman stated: "The night was clear though there was no moon, and the airship could be distinctly seen, and the whizzing of its motor was heard by us all." (8)

A traffic foreman confirmed: 'There is no question about the reality of the mysterious airship. Too many of us have seen it to leave room for doubt. We could not all be mistaken. The airship took a wide curve from the direction of Newport, and though high up could be clearly seen against the clear sky even if it had not been lit up by the two lights which it carried, and we all heard distinctly the whirr of its driving gear. It seemed to hover over the docks for a few seconds, and then swept away across the Channel, and the lights were extinguished as it passed away to the eastwards. We could not see those on board. The airship was too far up for that at night, but it was plain that it was a big airship with the usual cigar-shaped balloon." (9)

Residents in Salisbury Road, Cathays, Cardiff, also claimed they saw an airship on 18 May between 10.40 and 10.50PM but none of the neighborhood police on night duty saw any sign of it. (10)

Lethbridge's story was so sensational that reporters insisted on him taking them to the landing site. The area was so rugged that they abandoned their taxi and had to walk a quarter of a mile to reach the spot. Here they found a 54-foot long gouge in the ground as if made by a plowshare, the grass was trampled and news clippings referring to the German Army, airships and spies were scattered about. Bits of blue paper with figures and letters on them that were not "of the average English calligraphy" were also at the locality along with a lid of a tin of metal polish and clumps of papier-mâché paper.

A clue as to their origins was given by the cut-up letterhead of Arthur Shirley & Co., of Threadneedle Street, London, with the words "provincial centres...rest assured we shall not...the fullest confidence...this letter amply justified' typewritten on it. When contacted they were puzzled by how their letter got there although the head of firm did admit: "They are from a letter I have sent to several correspondents in Wales. I have several friends in Wales who have taken out airship patents, but I know nothing of this affair." (11)

There was also considerable excitement about a red label with French instructions on it, attached to a chain and a small plug or pin. One suggestion was that it came from a gas cylinder used for inflating car tyres or a fire extinguisher. A manager of the Michelin Tyre Company identified it as a valve-cap that is fitted to their car tyre inflators. French instructions were only supplied with Continental orders, and a passing motorist could not have dropped it as motor vehicles were unable to reach the landing spot. A spokesman for Michelin said: "We know by Mr. Lethbridge's statement that the airship in question is provided with a carriage, and we also know that earlier types of aerostats and aeroplanes were fitted with bicycle wheels to give them their first impetus. It would appear, therefore, that the mysterious airship was fitted with Michelin cycle tyres, all of which were provided with this type of valve." (12)

On May 18, 1909, Charles Lethbridge, while traveling near
the top of Caerphilly Mountain, South Wales, came across a
"long, tube-shaped affair" lying on the grass on the roadside. Nearby were two men wearing big, heavy fur-coats,
and fur-caps. Later, a red label with French instructions
on it was found at the landing site. Attached to the
label was a chain along with a small plug or pin.

"*The Evening Express*" tells of Mr. O. Riddervold, a Norwegian now in residence in Cardiff, who called at their office "to offer an explanation of the label found on Caerphilly Mountain on the spot where the mysterious airship is said to have rested." The report goes on to say:

"Mr. Riddervold has been engaged in airship construction in France and England, and he gave an exposition of the purpose to which the pin attached to the label is applied in the motor mechanism of an airship. It was Mr. Riddervold's conviction that the pin was the instrument used for releasing the valve fixed to the pump in order to inject air from the atmosphere into the balloonette of the airship. This balloonette is concealed within the body of the ship, and when the gas escapes from the latter compressed air is pumped into it from the balloonette, so that the canvas shall not sag and thus interfere with the control of the machine.

"Mr. Ridervold was convinced that the French word '*obus*' on the label, although meaning shell, was not intended to apply in any sense to a shrapnel shell. The pin, however, is worked on the same principle as the time-fuse of a military shell, and that is how the word 'obus' came to be applied to the motor mechanism of airships.

"Asked if the discovery of the label on the mountain was evidence that an airship had been there, Mr. Riddervold said it was undoubtedly testimony that an airship had either rested on that particular spot or had passed over it, and that the label had been dropped by the aviators." (13)

The Hon. C. S. Rolls, the well-known aeronaut, motorist and founder of the Aero Club, who went on to co-found Rolls Royce was asked for his opinion; His view was: "The whole thing is a mystery. There is either no airship at all or else it is a foreign one. At Cardiff there has been a dirigible balloon built, but it has been stated that it could not be the one. That being the case, I could not see how it could be an English machine. If this had been the ease, as some people suggest, we would have been bound to have heard of it before, because a dirigible balloon requires a very large shed for building, and could never have been filled without it becoming known very quickly.

"I see nothing impossible in a German airship coming across, because the new airships of the German Army have a range of 800 miles. The French have also airships which are capable of doing such distances; but I do not think there is any machine in England which is capable of doing such a distance, and no other Powers have. Therefore, it must either be French or German property."

When asked about the articles found on the ground at Caerphilly, he said: "No, I don't think they had anything to do with it." (14)

So who piloted this mystery airship? Paul Brodtman, the managing director of the Continental Tyre Company, was alleged to have launched a model airship and towed it with a motor car to experiment with the art of aerial advertisement.

DIALOGUE WITH AN EVASIVE GERMAN

"A '*Morning Leader*' representative found himself last evening in the Clerkenwell-road, London, E.C., with a clear sky overhead undimmed and unencumbered by airships (at present), and knocking at the mahogany door of the sanctum sanctorum of the advertisement manager of the Continental Tyre Company.

"A guttural voice - unmistakably German - said, 'Come in.'

"Entering, somewhat timidly; our correspondent found himself in the presence of two young men with alert eyes and laughter-wrinkles in their cheeks. One of them was Mr. Paul Brodtman, the managing director of the company.

"In reply to a series of questions, Mr. Brodtman spoke (and please note it) in guttural tones."

Brodtman acknowledged his company was one of the first to use large toy balloons for aerial advertising and that his showroom contained some cigar-shaped examples that were several feet long. And this dialogue ensued:

'Can they fly?'

'Yes, if they are...'

'Towed at the end of a line, with the other end fixed to a fast motor-car?'

'Quite so.'

'You are very fond of motoring, Mr. Brodtman; you possess a fast car?'

'A ripper,' smiled the manager. 'She can go.'

'And when the envelope is filled with oxygen (say), and the airship is in tow of a.car which is a ripper and can go, the propeller would whizz as the ship sped through the stilly night?'

'Of course - naturally.'

'And scare Punch and Judy show proprietors on the way home across the Welsh mountains, particularly if two fur-coated gentlemen were talking at the same time in guttural tones?'

'That would depend upon the state of nerves of the Punch and Judy man. And now I must really ask you to excuse me, as I have to catch the train to Liverpool. I really cannot say any more-at present. Indeed, I know nothing."

Just to underline his connection with the Caerphilly case it was noted that Brodtman's colleague "walked away, and left the room, taking with him a heavy fur coat that had been hanging (unobserved) on a peg near the door." (15)

Shortly after this revelation, Mr Brodtman vigorously denied any connection with the airship sightings and claimed he was totally misrepresented. (16)

HOAXES? STUNTS? DECEPTIONS?

Percival Spencer, a well-known aeronaut and airship constructor, revealed that he had sold several twenty-five feet long model airships. These used a small lamp to generate heat to keep them aloft, which might explain why people always saw a "searchlight" coming from the airship. He had also sold five large man-carrying airships, though none of them were attributed to Lethbridge's sighting.

The idea of a secret inventor was also discussed. A Dr. M.B. Boyd came forward to say that he had spent eight years perfecting an airship that was 120 feet long and capable of traveling 1,000 miles non-stop. It carried two wings, a cabin that could carry three crew that was integrated into the envelope and had three sets of wheels that enabled it to be used like a motor car on the ground. The craft was kept in a secret shed only one hour's drive from London. He said his craft was responsible for Lethbridge's sighting and for the sightings in Ireland. Needless to say no more was ever heard of this revolutionary aircraft. (17)

On the night of 24 May the lights and sound of an airship were heard by a night duty worker at the Sewell lime works. The next morning the wreckage of an airship was found in a hedgerow at Sewell and crowds of people from nearby Dunstable, Bedfordshire, came to look at this marvel. It was explained that this was a bamboo-framed unmanned airship that carried a motor and lights, and had been tossed from the sky by stormy weather. A note amongst the wreckage promised a £5 reward for its return to "a West End firm of motor factors." (18)

A reporter from the "*Manchester Daily Dispatch*" was skeptical that this airship had actually taken to the air.

"It is certain that such an airship, even when fitted together and the missing parts supplied, would not carry a man and it is questionable if this one ever flew at all. The conclusion one is bound to come to is that the various parts that go to make up the 'airship' were taken to the spot where they were found and left there for some credulous and fearful person to discover." (19)

A Continental motor car manufacturer (Mr Brodtman's company?) was also blamed by The Standard, 26 May 1909, for the sightings. It explained that: "Two motor-cars were used; one carrying the balloon, a 20-h.p. motor, and a lot of bamboo poles for the steerable car, and the other half a dozen cylinders of compressed hydrogen. The first 'ascent' was on the Chelmsford road, at Writtle. The balloon was secured with ropes and held to the wheel of one of the motor cars. Lights were put out or darkened, watchers were told off to give an alarm on the approach of strangers, and whilst the 'steerable' was captive, toy fire-balloons were sent up to windward of it, and the engine of one of the motor cars was set working, the 'silencer' being opened to increase the noise in order to give the impression to any nocturnal observer whose eye might be attracted by the balloon that he heard the whirring of its motor overhead.'

Was this all a series of disconnected hoaxes or publicity stunts that caused so many sightings?

Did Mr. Brodtman use his ripper of a motor car to tow an airship up Caerphilly Mountain? If so that does not explain the men seen getting inside the craft or the debris found at the site.

STICKING TO HIS STORY

Was Lethbridge himself involved in a separate hoax? Lethbridge did work at Cardiff Docks in the winter but it seems a stretch that he was able to organize so many people to say they saw an airship the same night. And what about the gouge in the ground and the odd scattering of material at the site? Was that a scene set by the men observed on the mountain two days earlier? What made it such a strong case was that the newspaper reporters found Lethbridge to be "an elderly man, of quiet demeanour, and did not strike one as given to romancing." (20)

In July 1909 a representative of the *"Evening News"* re-visited Lethbridge and he still kept to his original story: "I am quite positive, however, that it was an airship I saw that evening as I was trundling my truck along the road over the mountain The night was a bit dark, but I distinctly saw the object rise from the ground in front of me and fly away in the direction of Cardiff, after two men had jumped into it. What I thought were rockers upon which the airship was resting on the ground now appear to have been the wheels on which it was carried along after it came to earth. I am not a practical man in this respect, and, of course, cannot enter into the details from a scientific point of view.

"You ask me whether I have been chaffed over the matter. Why, I should think so. I cannot go to the docks looking for work but I am assailed right and left, and I am sick of the whole matter, although I take all the badgering in good part.

"You know. I am a workman at the docks, and when there is no employment to be got there, I go about with my Punch and Judy show by invitation. I do not know how they may feel now, but down at the docks it has been extremely funny to me to hear the remarks passed as I walked along. 'Our airship is all the go again,' says one, and from another quarter the finger of scorn has been pointed at me as if I had been 'boozed.' Why, I don't drink to excess on any occasion, and I only had a sleever that night before I crossed the Caerphilly Mountain. Coming back, however, to the main part, I say that Dr. Boyd's story of his invention and his experiments bear out in their entirety my statement of what I saw on that evening, and I will not forget it." (21)

As a way of showing ufology incorporates the old with the new, this headline was posted above a short account of the Lethbridge story on Reddit in February 2022:

ALIEN ARTIFACTS

"On May 18, 1909 a UFO landed in Caerphilly Mount (sic), Wales piloted by two men in fur coats. It departed in the same zig-zagging fashion as the now well-known movement of the 'Tic Tac' as described by Cmdr. David Fravor."

References

1. Daily Express, 18 May 1909. East Coast Illustrated News (Clacton), 22 May 1909.

2. Daily Express, 18 May 1909.

3. Grove, Carl. "The Airship Wave of 1909?", FSR,16,6. Evening News, 15 May 1909.

East Anglian Daily Times, 18 May 1909. Irish News, 17 May 1909.

4. East Essex Advertiser (Clacton), 22 May 1909.

5. South Wales Daily News, 21 May 1909.

6. Cardiff Evening Express, 19 May 1909; South Wales Daily News, Western Mail, 20 May 1909.

7. South Wales Echo (Cardiff), 19 May 1909.

8. Western Mail (Cardiff), 20 May 1909.

9. South Wales Daily News, 20 May 1909.

10. ibid

11. Daily Express, Birmingham Gazette and Express, 20 May 1909.

12. Grimsby News, 25 May 1909.

13. Evening Express, 21 May 1909.

14. ibid.

15. ibid.

16. Evening Standard, 21 May 1909.

17. Daily News (London), 6 July 1909; The Aero, 13 July 1909.

18. 'Airship wreck, 25-5-1909' at:

https://dunstablehistory.co.uk/A_to_E/pages/Airship%20wreck,%2025-5-1909_jpg.htm

1909 Airship Crash in Sewell May 25th, at:

https://hrhsarchive.org.uk/archive/sewell/1232336-1909-airship-crash-in-sewell-may-25th

19. Manchester Daily Dispatch, 26 May 1909.

20. Cardiff Evening Express, 19 May 1909.

21. Evening News, 09 July 1909.

Further References

Charles Fort 'Lo!' Claude Kendall, 1931, pp 134-136. At: http://www.resologist.net/lo111.htm

Fort misspells his name 'Lithbridge' not Lethbridge.

Nigel Watson (ed.), 'TheScareship Mystery: A Survey of Phantom Airship Scares 1909-1918', Domra, 2000.

Nigel Watson 'UFOs of the First War: Phantom Airships, Balloons, Aircraft and Other Mysterious Aerial Phenomena', History Press, 2015.

Airminded website at:

http://airminded.org/archives/scareships-1909/

Research credit to John Hind, Granville Oldroyd, David Clarke and Dirk van der Werff.

The metal fragments from a UFO that was allegedly seen to explode near the Brazilian town of Ubatuba are still considered one of the most controversial of all physical-evidence cases.

4.

THE GRANDDADDY OF UFO DEBRIS ANALYSIS
By Tim R. Swartz

Those who have a casual interest in the subject of UFOs may think that the government probably has multiple containers full of crashed UFO debris, (and maybe an alien body or two floating in tanks) hidden away at the Pentagon, or Area 51.

Nothing could be further from the truth. Even though strange bits of metal and other strange substances, allegedly seen coming off of UFOs, have been found and collected for years, precious little has ended up in the hands of scientists to be properly assessed and analyzed.

This is what makes the Ubatuba, Brazil, incident so compelling, even if there are some details that are missing from the story. Not only were the pieces of metal put in the hands of experts fairly early on in the original investigation, new research, conducted in 2017 and 2018, seems to confirm that these metal fragments are indeed unusual.

THE APRO BULLETIN

One of the earliest accounts of this story published in the U.S. came from the March, 1960, issue of the Aerial Phenomena Research Organization (APRO) Bulletin. The article details that on September 14, 1957, Mr. Ibrahim Sued, a social columnist for the Rio de Janeiro daily newspaper "*o Globo*," featured in his column an unusual letter that he had received from a reader.

"Dear Mr. Ibrahim Sued. As a faithful reader of your column, and an admirer of yours, I wish to give you something of the highest interest to a newspaper man, concerning the flying saucers. If you believe they are real, of course. I also didn't believe anything said or published about them. But just a few days ago I had to change my mind. I was fishing together with some friends at a place near the town of Ubatuba, Sao Paulo, when I saw a flying disk. It approached the beach at unbelievable speed, an accident seeming imminent – in other words, a crash into the sea. At the last moment, however, when it was about to strike the water, it made a sharp turn upwards and climbed up rapidly in a fantastic maneuver. We followed the spectacle with our eyes, startled, when we saw the disk explode in flames. It disintegrated into thousands of fiery fragments, which fell sparkling with magnificent brightness. They looked like fireworks, in spite of the time of the accident - at noon.

"Most of these fragments, almost all, fell into the sea. But a number of small pieces fell close to the beach and we picked up a large amount of this material – which was as light as paper. I enclose herewith a small sample of it. I don't know anyone that could be trusted to whom I might send it for analysis. I never read about a flying saucer having been found, or about fragments or parts of a saucer that had been picked up; unless it had been done by military authorities and the whole thing kept as a top-secret subject. I am certain that the matter will be of great interest to the brilliant columnist and I am sending two copies of this letter to the newspaper and to your home."

Unfortunately the signature was illegible and the writer was never identified.

Dr. Olavo T. Fontes, APRO's Brazilian representative, read the letter in the column and decided to call Mr. Sued to see if he could have a look at the material. Shortly thereafter, Dr. Fontes was invited to Sued's home and allowed to examine the metal.

The pieces were dull grey, solid and appeared to be metallic. They were rough and irregular, with scattered whitish areas on the surface produced by the deposit of a thin layer of a powdered substance which could easily be removed with a fingernail. Fontes noted that the stuff was lighter than aluminum – almost as light as paper.

SPECTROGRAPHIC ANALYSIS SHOWS SURPRISING RESULTS

Dr. Fontes told Sued that he had friends in the scientific community that could examine the material. Sued, who said that he wasn't particularly interested in the subject of UFOs, agreed, as long as he was kept informed about the results.

Dr. Fontes submitted a part of the sample to the Mineral Production Laboratory, a division of the National Department of Mineral Production, a Brazilian government lab. Fontes was introduced to Dr. Pfeigell, the chief chemist, by a friend.

Pfeigell was at the time engaged in special work with plastics and turned it over to Dr. David Goldscheim, one of his assistants who, after studying the material, said they could be the fragments of a meteorite...Dr Pfeigell didn't agree, because of the light weight of the substance, and personally conducted a test using phosphomolydic acid to positively determine that the substance was indeed metal.

He then decided on a spectrographic analysis with the official analysis of the substance conducted by chief chemist of the Spectrographic Section of the Mineral Production Laboratory, Dr. Luisa Maria A. Barbosa.

Her report, which was dated "September 24, 1957," noted that the sample received included two fragments of metallic appearance, grey color, low density, and weighing, each one, approximately 0.6 gr. The spectrographic analysis showed the presence of magnesium (Mg) of a high degree of purity and the absence of any other metallic element.

Another test, conducted by the Laboratory of Crystallography at the Geology and Mineralogy Division of the National Department of Mineral Production, using X-Ray diffraction, showed that the substance was magnesium and apparently absolutely pure.

A Geiger counter and an Atomic Scaler were also used to determine whether the fragments registered any extraordinary amount of radiation. No abnormal amount was found.

THE DEBUNKING COMMENCES

Of course skeptics had a field day trashing the alleged "metal from outer space." The fact that the fragments were sent in anonymously convinced many that this was nothing more than a hoax. Others pointed out that Dr. Fontes visited Ubatuba sometime between 1957 and 1960 but was unable to find anyone who saw a UFO explode over the beach. Something that should have garnered considerable attention if it happened around noon as the anonymous writer claimed.

Further muddying the waters on the Ubatuba material, the owners of the Hotel Casarão da Lagoinha Hotel, near Ubatuba, told investigators that some time in the 1930s, fishermen in the area had observed a bright light and an explosion in the sky. Afterwards, while fishing off Anchieta Island, they picked up a "strange stone" in their net. The stone was about eight inches and extremely light in weight. It was later handed over to the director of the prison on Anchieta Island, Major Nilton Feliciano dos Santos. Allegedly, Major dos Santos took the rock to a laboratory in Sao Paulo city for analysis, and it was found to have an "unusual" composition.

Whether or not this story has any connection to the metal sent to the "*o Globo*" in 1957 is debatable. It does seem to be somewhat suspicious that this area has several stories about strange lights and explosions in the sky dating back several decades.

NEW TECHNIQUES FOR ANALYSIS

However, despite the fact that the origins of the metal fragments have been obscured, modern techniques for a closer look at the composition of the material have been used with interesting results.

A 2022 paper called "*Isotope Ratios and Chemical Analysis of the 1957 Brazilian Ubatuba Fragment*," by Robert M. Powell, was published in the "*Journal of Scientific Exploration Anomalistics and Frontier Science*" (Vol. 36 No.

Ubatuba is located on the southeast coast of Brazil in the State of São Paulo.

1). The author details that recent tests completed in 2017 and 2018 indicated that the debris was mostly composed of extremely pure magnesium with an odd strontium impurity. This formula was not used in the manufacture of magnesium in the 1950s.

A sample from the Ubatuba fragment was tested with the intent of examining the isotope ratios of its primary element, magnesium, and the trace elements strontium, barium, copper, and zinc. For the first time, HR-ICPMS (High Resolution Inductively Coupled Plasma Mass Spectrometer) techniques were used to look at the isotopic ratios of the minor constituents as well as the primary magnesium component of the sample. The magnesium isotope ratios were found to fall within terrestrial limits while the results on the isotope ratios of the trace elements were inconclusive.

Nevertheless, one strange aspect of the Ubatuba sample does remain. All testing consistently indicates that the Ubatuba sample is 99.88% pure magnesium with traces of strontium, barium, zinc, and copper. The strontium impurity is not a normal by-product in the manufacture of magnesium and would have been intentionally added.

Apparently, when shown the results of the tests, a Dr. Beaman and Dr. Solaski of DOW Chemical, two men very familiar with the production of magnesium, were surprised by the presence of strontium and couldn't figure out why it was in the sample.

These new results on alleged UFO debris collected more than 60 years ago show that analyzing strange alien artifacts is still not a perfect science. It still leaves us with the mystery as to how high purity magnesium with the addition of strontium impurities showed up at a Brazilian newspaper office in 1957.

ALIEN ARTIFACTS

Strange Substances Falls On Ranch In Washington

Larry Robinson of Sequim, Washington, found an unkown object on May 2, 1963 which apparently fell from the sky.

He made the discovery on his ranch which is located on Sherbourne Road near Sequim, Washington. Several pieces of the "thing" were found in a corral. The largest piece measured about 12 inches long, six inches wide and three inches thick. It appeared to be lava-like and porous. Its color is described as grey, but the pieces were covered on the outside by a white powder.

Tiny bits of crystal appeared underneath the powder. When tasted, the stuff had a salty taste.

A local science teacher, Mrs. James Scott, felt that after microscopic examination the object was not a meteorite, primarily because of the deterioration of the white powder. A piece of the stuff was sent to the University of Washington Geology department for analysis and no results have been announced concerning their findings.

APRO Bulletin - September, 1963

5.

THE ARTIFACTS OF HIGH STRANGENESS
By Scott Corrales

Regardless of your cultural background, it is almost impossible not to have heard about the magical artifacts associated with the past—in myth and in fiction. These range from the swords of King Arthur — Excalibur— and Roland— Durandana—to the rings of the Nibelungen (in Wagner's "Gotterdammerung") and the ring of Sauron (in Tolkien's "Lord of the Rings"), and can even include semi-mythical items such as the staff with which Moses turned the waters of the Nile into blood before the startled eyes of Pharaoh and the military banner that preceded Emperor Constantine into victory, and upon which was written "In Hoc Signo Vinces."Can the Wicked Witch's "mirror, mirror on the wall..." in Snow White be a veiled reference to the sorcerer John Dee's seeing stone, or to the more ominous magic mirror of the Aztecs?

FALLEN OBJECTS POSSESSING GREAT POWERS: IS THERE AN ET SOURCE?

Many traditions speak of objects having fallen from the sky and conferring immense power, such as the Kaaba stone, which exists to this very day. Ancient kings in the Middle East would loot the tombs of their forebears in hopes of acquiring weapons and armor which would enhance their own prestige and kindle the valor of their armies. What greater stimulus than, say, appearing before your legions wearing a breastplate allegedly taken from the lost tomb of Alexander the Great (stolen by Caligula from the hero's tomb in Alexandria, according to Roman historian Dio Cassius), or a blade once wielded by the legendary Persian hero Rustam?

Few people outside of South America have ever heard of this most mysterious and controversial emblem of power, which according to some sources, may be the ultimate source of mysteries.

Tradition holds that the Baton of Command (a direct translation of its Spanish name, Bastón de Mando, which in turn translates as Simihuinqui — the name given to it by the South American tribesmen) was crafted some eight thousand years ago by Multán (also known as Voltán), a mighty chieftain of the Comechingones tribe, from a piece of black basalt. The occult powers of this ancient artifact were legendary among the tribes of the modern Argentinean Chaco and the Bolivian lowlands, and in the1830s, an Araucanian warlord named Calfucurá—well steeped in his people's traditions—led a massive search for the object in the mountain ranges of Tandil, Balrcarce, San Luis and Córdoba which did not turn up the Baton of Command.

It is at this point that we must delve into the other esoteric tradition linked to this black basalt wand: students of the occult believe that aside from its Neolithic age, the Baton of Command is tied in to the European tradition of the Holy Grail, which has been handed down to us through Arthurian legend and Wagnerian opera and is far removed from fiction.

These esoterics, like the late Argentinean scholar Guillermo Terrera, believe that the 12th century chansons de geste of Chretien de Troyes and Wolfram Von Eschenbach make allusions to the Baton of Command and to the existence of South America—a landmass whose existence medieval man could not have suspected. While these allegations would quite rightly be dismissed as crankery in the hallowed halls of academe, Terrera and his followers nevertheless make an intriguing case for their beliefs.

WHAT IS THE 'STONE OF WISDOM'?

According to these esoteric revisionists, mythological sources in Central and Eastern Asia make reference to a mysterious character entrusted with the custody of two sacred items: one of them the so called Holy Grail of Sangraal, and the other being "the Stone of Wisdom," which they identify as the Baton of Command.

The enigmatic custodian of these items would have begun his career thousands of years ago, and is only known as the "Man from Persia" — the Parsifal of Eschenbach's songs, and the Sir Perceval of the Arthurian Cycle. According to the German minstrel's epic, the enigmatic Parsifal travelled to the land of Argentum ("...the secret gates of a silent land named Argentum and will always be...") to lay the objects under his care in the sacred hill known as Vlarava. Extrapolating from the epic poem, these esotericists have identified Argentum with Argentina and the sacred mount Vlarava with Mount Uritorco in the country's northern reaches.

Putting aside their reliance on the late medieval epic for a moment, Terrera and his colleagues further noted that the knighthood of the Grail mentioned in the songs is none other than that of the Knights Templar, about whom much has already been written. Their belief is borne out by the fact that the Templars seemed obsessed with recovering a holy relic which was variously known as the "Stone of Wisdom" or the "Talking Stone." Could this have been the Baton of Command?

In 1934, a mystic named Orfelio Ulises, who had just returned to Argentina after having spent eight years in Tibet as an adept of Lamaism, came upon the mysterious Baton of Command, allegedly "guided" by the mental powers of his Tibetan masters, and dug the object out of the slopes of Mount Uritorco in Capilla del Monte. While all of this smacks of Madame Blavatsky in all her glory, other more credible events would also come to pass.

KEEPING THE BATON HIDDEN FROM THE NAZIS

Much like Spielberg's Indiana Jones, Ulises would come to realize that other parties were interested in his discovery: The Nazi Ahnenerbe ("Ancestral Heritage Society"), founded by Heinrich Himmler in 1935 with the aim of supporting the theories put forth by the notorious Thule Society, had already secured paranormal objects like the Spear of Longinus—also known as the Spear of Destiny—in 1938, and a year earlier had started to send out worldwide expeditions in search of Noah's Ark, Atlantis, and bizarre medicines used by South American natives. It was only a matter of time before these twisted forces had fixed their predatory gaze on the Baton of Command. To their aid came then-colonel Juan Domingo Perón--Argentina's future dictator. Perón spent the late

1930s as a military observer in Italy and Germany and was also fascinated by the occult.

Orfelio Ulises and a number of "hermetic scholars" managed to conceal the periapt from the Nazis and keep it in Argentina, where it remained under Ulises' care until his death, and then passed on to Professor Guillermo Terrera in 1948. It is currently in the custody of Dr. Fernando Fluguerto Martí and his Delphos Group.

Also in 1948, Baron Georg Von Hauenschild, an archaeologist and Grail scholar, prepared an exhaustive report on the Baton of Command for the Institute of Archaeology, Linguistics and Folklore of the University of Cordoba, showing that the objects estimated age was indeed 8000 years and of clearly Neolithic manufacture. Great care was taken by prehistoric craftsmen in polishing the object, rounding off its base and tapering its head into a soft conical shape. The volcanic basalt that it is made of gives it a metallic look, and when struck, the Baton of Command makes a ringing sound. Subsequent electromagnetic and spectroscopic analyses proved that the Baton emits an electromagnetic field; students of the occult have construed this to mean that a properly trained adept, under the right conditions, might be able to establish a paraphysical link to other realities or unlock the wand's secrets.

This is where the Baton of Command's powers apparently lie: it was designed, according to Professor Terrera, as a means of regenerating humanity and patiently awaits the right person to come and make use of it. As of this writing, that person has apparently not come.

Author Luis Alberto Vence makes the following curious note. According to historical sources and the beliefs of contemporary smiths and armorers, the mythical blade Excalibur would have measured approximately 1.10 meters — the exact length of the Simihuinqui or Baton of Command.

Metaphysical claptrap or occult truth? You be the judge. In his book *"El Valle de losEspíritus"* (Buenos Aires, Kier, 1989) Terrera sums up the situation thus: "We must bear in mind that all that science has discovered up until yesterday as an absolute truth could be corrected either today or tomorrow, since all human knowledge is subject to change, as part of the dynamic process that accompanies it."

THE POWERS OF SACRED JEWELS

In 1997, moviegoers were treated to John Cameron's *"Titanic"* and its subplot concerning an intriguing blue diamond. Jewels such as the one shown in the film have often been ascribed remarkable talismanic powers, and in other cases, qualities that make them lethal to the user, much like the One Ring in J.R.R. Tolkien's saga.

Not many of these items have survived down to our time, but we know that Alexander the Great was particularly fond of an unusual opal which kept him from being wounded in battle. Upon embarking on his conquest of the Persian Empire, the Macedonian king (whose own armor would become talismanic over the centuries, as mentioned earlier) made a quick stop at the ruins of Troy to secure a sacred shield, which had belonged to one of the heroes of Homeric legend, in an effort to bolster his invulnerability an extra notch. But neither the exotic opal nor the ancient shield were much help when an arrow pierced Alexander's lungs while storming the walls of a city in the Punjab.

Rings occupy a privileged position among all articles of jewelry: Apollonius of Tyana received a ring of amber from one of the initiates in the fabled city of Iarchas somewhere in Central Asia (or another dimension?). The amber stone allegedly kept its wearer from harm and enabled him or her to have foreknowledge of any dangers ahead—a faculty that the legendary Apollonius employed more than once. Charlemagne possessed an unusual ring whose stone was supposed to preserve all of a warlord's conquests. Naturally, this amulet was quickly taken by the Frankish monarch's son Lewis and in turn squabbled over by Charlemagne's nephews, who divided up their grandfather's empire.

But the powers ascribed to these adornments mainly reflect the wishes of the human wearer rather than any true supernatural powers. However, what are we to make of the ring worn by Charles XII of Sweden? This Scandinavian monarch ruled an empire built around the Baltic Sea and was one of Russia's most implacable foes. Author Brad Steiger notes in his book *"Atlantis Rising"* (Dell, 1976) that the Swedish king's rise to power had apparently been aided and abetted by his dealings with a "little grey man" who had given him a ring that would vanish on the day of Charles' death. The monarch appears to have gladly accepted this gift and embarked on his military career. In the heat of battle, shortly after one of his officers noted that the ring had vanished from his fingers, the monarch received a mortal wound.

Emeralds held a particular fascination for the infamous Emperor Nero, according to the historian Pliny, who wrote that the lyre-strumming despot owned a flat, nameless specimen imbued with supernatural powers, which he even used as a magnifying glass. While antiquity was fascinated by colored stones like sapphires and rubies, diamonds acquired importance in more recent centuries—some of them having names and histories as bizarre as any fictional object, and the Hope, Star of India and Kohinoor diamonds have been featured on silver screen.

The Regent diamond is one of the more fascinating ones. A slave in an Indian mine found the precious stone sometime during the 1600s and escaped bondage, only to be slain by a sailor to whom he had shown the diamond. The sailor took the stone to France, where he died a suicide. The Regent changed hands from one French aristocrat to the next, bringing misfortune to all of them. Napoleon Bonaparte had the Regent embedded in the pommel of his sword, which he later surrendered upon being exiled to Elba in 1814.

RELICS OF THE CRUCIFIXION

There are two sacred objects which occupy positions of honor in Western tradition. One is allegedly on display in an Austrian museum, while the other is still avidly sought by occultists and adventurers to this very day: the former is the Spear of Longinus and the latter is the Holy Grail or Sangreal.

In the year 1099, as the warriors of the First Crusade lay siege to the mighty city of Antioch in what is now modern Syria, a priest named Peter Bartholomy presented himself before the expedition's leaders. Saint Andrew, he said, had appeared to him in a dream, ordering him to rescue "the steel head of the lance that pierced the side of our Redeemer" and which lay in the ground beneath the altar of the Church of St.

Peter. The heavenly visitor further commanded that the Spear should be carried into battle against the enemy, "and that mystic weapon shall penetrate the souls of the miscreants."

The Crusader bishops scoffed at the suggestion, but Count Raymond of Toulouse, one of the expedition's leaders, saw that the morale of his knights was ebbing in the Middle Eastern heat. Laborers dug under the altar to a depth of twelve feet; Peter Bartholomy descended into the hole and retrieved the spear head, which was welcomed with great reverence and wrapped in a cloth of gold. The sacred relic rekindled the martial spirit of the Crusaders and led them to victory.

However, the historian Edward Gibbon points out that the relic was in fact a fraud. Bartholomy had in fact retrieved the point of a Moslem spear which had been secreted in the pit the night before his descent to find it. "The holy lance," notes Gibbon, "soon vanished in contempt and oblivion."

Or did it?

Trevor Ravenscroft, author of the highly controversial *"The Spear of Destiny"* (Weiser, 1973), charts the path of the holy lance as it changes hands from owner to owner: originally forged by the prophet Phineas as a symbol of the magical powers of God's Chosen, it becomes the Spear of Herod and is seized by the Roman centurion Longinus to pierce Jesus on the cross, and from there on is wielded by every great hero in western tradition: Constantine the Great (who apparently used it to trace out the borders of his new city on the Bosphorus), the chieftain Alaric, the Byzantine emperor Justinian, and Charlemagne, to name only a few.

Whether we are inclined to side with Ravenscroft or not on this point, the author correctly mentions the existence of many such "holy lances" in the Christian tradition: one of them in the Vatican, another in Cracow, and another on display in Vienna, which was allegedly brought to France by the Crusader king, St. Louis. Could this latter have been the fraudulent spear deposited by Peter Bartholomy in Antioch in the year 1099?

According to legend, the Holy Lance holds sacred powers and the person who possesses it becomes invincible and capable of ruling the world. The Lance is on display at the Imperial Treasury in the Imperial Palace in Vienna.

The veneration of such relics, ranging from the bones and hair of certain saints to objects like the holy lance and fragments of the True Cross, was a major feature of medieval Christianity and the source of much pilgrimage in those troubled times. Many of these relics were manufactured as tourist attractions, so to speak, and have cast doubt upon the credibility of the rest, as evidenced by the dispute raging over the Shroud of Turin. The holy lance, then, remains tantalizingly within our grasp, but forever out of reach.

A KING FOR THREE THOUSAND YEARS

God or human wizard? All books of esoteric lore speak reverently of Hermes Trismegistus or Hermes Thrice Great and his coveted "Emerald Tablet." Worshipped by the Greek residents of the Egyptian city of Alexandria, and identified with the ancient deity Thoth, the scribe of the underworld, Hermes Trismegistus was believed to have been a human monarch who ruled for three thousand years and wrote an amazing thirty-five thousand books — a useful way of filling up three millennia. Yet only fragments of this mythic figure's writings have been handed down from hoary antiquity, ironically through the works of Christian authors.

Although some modern scholars agree that Hermes Thrice Great was in fact the title given to the proto-chemist in charge of refining gold—a seemingly "magical" process to the ancients—medieval alchemists and thinkers considered Trismegistus to have handed down secrets preserved by the aptly-named "hermetic" schools of knowledge.

The most significant of these works was a document referred to as the Emerald Tablet, which was supposedly buried along with Trismegistus's mummy under the Great

Pyramid of Gizeh. The Tablet allegedly reveals the secrets of alchemy. Although the Hermes Thrice Great's mummy still waits patiently for archaeologists to find it (although the "Tomb of Osiris" discovered in 1998 does offer fascinating possibilities), part of the Emerald Tablet's metallurgical secrets can be found in the Leyden Papyrus—brought back to Europe in the 1820s by Johann d'Anastasi—which escaped the destruction of alchemical texts mandated by the Emperor Diocletian in 298 A.D..

A FORBIDDEN BOOK

Based on this historical assessment, one could hardly consider the Emerald Table a holy relic...unless the theories of Argentinean author Fabio Zerpa are taken into consideration.

Zerpa, better known for his work in ufology, cites the Count de Gebélin's belief that the Emerald Tablet is merely another name for the legendary Book of Thoth — a forbidden book some ten thousand years old which would have been the basis of Egyptian civilization and occultism, as well as the key to "mastering the secrets of the air, the sea, the earth and the heavenly bodies." In *Primitive World,*" his treatise on Egypt, de Gebélin remarks that the Book of Thoth survived destruction because it was cleverly disguised as a game, as we shall see below.

An Egyptian priest, Nefer-Ka-Ptah, retrieved the book, which had been sealed in a series of nested sarcophagi and kept in the bottom of Nile. Upon studying it, the priest was able to learn the art of numerology, communication with entities living across space and time, clairvoyance, and the art of building "magic mirrors" which do not reflect the viewer's countenance, but rather other worlds inhabited by loathsome beings.

Nefer-Ka-Ptah died a suicide, according to the story, and the Book of Thoth was spirited out of Egypt. Its magical powers and hidden knowledge would spread around the world in the form of the Minor and Major Arcana of the Tarot, which first appeared around 1200 A.D. in Italy as carticellas ("little cards") and were banned in 1240 and 1329 by bishops across Europe as malign. In his book "*The Black Art*" (Paperback Library, 1968) Rollo Ahmed, notes that the High Priestess card represents the Egyptian goddess Isis— perhaps the most tangible link to its Egyptian origin.

So, if Zerpa is right, the Tarot deck in your drawer could harken back to mythological times, placing it among the oldest relics known to mankind.

THE MULTIFACETED DR. JOHN DEE

Mathematician, astrologer, alchemist, spy, close advisor to queens and emperors: these are the impressive credentials of Dr. John Dee, one of the 16th century's most influential personages. Although he is best remembered for his work in the esoteric arts, mainly the development of the Enochian language employed in magical rituals, it is possible to find endless references to Dee's importance as political and scientific figure without a single mention of the aspects which have made him a household name in occult circles.

John Dee's achievements in esoterica—alleged communication with an order of angelic beings—were achieved through the technique known as "scrying," looking into mirrors or similar reflective surfaces such as bowls filled with water, mercury or oil, in order to have clairvoyant experiences. Dr. Dee himself lacked this ability, and depended on his assistant Edward Kelley to do the viewing (a technique very similar to a modern-day Remote Viewer and his handler). The techniques involved in the process of speaking to otherworldly entities are contained in Dee's "Libri Mysteriorum."

The reflective surfaces employed in the scrying were a globe of rock crystal--a precursor of the "crystal ball"—and a flat surface which Dee referred to as his "jet shewstone." These items are important relics of the paranormal tradition and survive to this very day, currently displayed in the British Museum.

Where Dr. Dee acquired his objects of power is a mystery. Nevertheless, there has been the suggestion— posited by paranormal researcher and playwright Eugenia Macer-Story--that the good doctor may have obtained them, by means of the activities of English "seadogs" raiding Spanish galleons, from the place they were most available at the time: Aztec Mexico, only recently conquered by Spain. The Aztec priesthood had fashioned a great many magic mirrors out of obsidian, and some of them are in museums, like the legendary black mirror of the evil deity Tezcatlipoca, on display in the Mexico City's Instituto Nacional de Antropología.

A MACHINE FOR SPEAKING TO THE GODS

There exist other objects allegedly employed for the purpose of communication with other levels of existence. One of them leads us into a discussion of the ever-controversial Knights Templar, the monastic order of warriors whose activities had a major impact on Europe and the Mediterranean Basin during almost two centuries. The Templars are perhaps better known for their activities during the Crusades and the tragic end of their order at the hands of the kings of France, but a number of scholars have focused on the occult aspects of their work. George Andrews cites French paranormalist Guy Tarade's research into a document dating back to the year 1310, which contains the "transcript" of the torture of Knights Templar by Church authorities. The tormented warrior-monk speaks of time travel, fiery chariots, wells of darkness in the heavens and realms of existence around unknown stars. Logically, this can be dismissed as pain-induced delirium, but the transcript hints at these things being seen through a "chest made of an unknown metal" tentatively identified with the Ark of the Covenant.

Here we take another flying leap into speculation: aside from all the powers ascribed to it over the millennia, could the Ark have been a means of seeing into other places and times? Andrews suggests that the "well of darkness in the heavens" is an unspecialized description of the astronomical phenomenon our scientists term a Black Hole—something utterly unknown in the 14th century.

It is with some trepidation that any writer approaches the subject of the Ark, since theories about its nature branch out like the leaves of a tree into unsuspected directions, making a cursory examination nearly impossible. In the limited space available to us here, we shall try to examine some of the most provocative thoughts on this, the most spoken-of relic that is out of our hands.

Viewers of Steven Spielberg's *"Raiders of the Lost Ark"* already know the basics: the Ark was a transportable device given by Yahweh to the ancient Israelites as a means of communication and occasionally as a weapon. The holy object was stored in the Temple of Jerusalem where presumably only members of the priesthood had access to it, and was kept safe from capture during the various invasions of Palestine by foreign powers (Egyptians, Assyrians and Hellenic Syrians). Although the Emperor Titus successfully conquered Jerusalem in 70 A.D., his triumphal arch in Rome, which shows

Roman legionaries on parade with their captured booty from the temple (the Menorah, sacred trumpets and tables), does not include the Ark—a sculptor's oversight, perhaps? These objects remained in Rome until the city was sacked by the marauding Vandals in the 5th century and taken to their capital, Carthage. The Byzantine armies of Belisarius shipped the objects to Constantinople after the conquest of the Vandal kingdom, but the superstitious Emperor Justinian, fearing that the captured "treasure of the Jews" would spell the ruin of Constantinople, had the objects sent to Jerusalem in 555 A.D..

A CASTLE TO HIDE THE ARK IN

Modern writers of occult history suggest that the Knights Templar discovered the Ark in the ruins of Solomon's temple and took custody of it, eventually shipping it back to Europe. A number of hiding places have been suggested for it: one of them is Rennes le Chateau in France, certain European forests and even remote Abyssinia. Some authors have raised the possibility that before reaching its ultimate resting place, the Ark may have been guarded in a very unusual location: the fortress known as Castel del Monte, located in the "heel" of the boot-shaped Italian peninsula.

Castel del Monte is located in Andria, a small town in the southern region of Puglia, Italy. It has been suggested that that the castle could have been a temple of knowledge where scientists were free to study undisturbed.

Castel del Monte was built in 1240 A.D. at the command of Frederick II, holder of an impressive number of titles, including Holy Roman Emperor and King of Jerusalem. A patron and ally of the Knights Templar, the emperor decreed that his strange, octagonal castle be built to precise measurements having magical significance and enclosing a main hall known, suggestively, as the Master's Chamber. The late Robert Charroux suggested that Castel del Monte was meant to be "a castle of Templar alchemists, governed by the figure 8, which when written horizontally, is the symbol of infinity and universal domination." (Charroux, *"Legacy of the Gods,"* NY: Berkeley, 1974).

Lacking all the typical inner structures of a castle, such as armories, refectories and living quarters, this octagonal fortress was not meant to repel invaders or serve as a garrison. In the light of all of its mystical associations, could we not speculate that this, in fact, was the special place built to receive the ultimate relic—the Ark of the Covenant? Under the protection of the powerful German emperor and the Knights Templar, it is hard to conceive of a safer location, or, as Charroux points out, a more symbolic one, since Castel del Monte is located halfway from the greatest points of pilgrimage in the Mediterranean world: Santiago de Compostela in the west and Jerusalem in the east.

Objects of such mystical prowess often conferred legitimacy upon the wearer: the crown of Constantine hung in full view above the altar of Constantinople's Hagia Sophia church, from where it was taken many times by anyone inclined to make a bid for the Byzantine throne. The successful coup-de-etat was seen as a sign of divine favor and the crown returned to its proper place.

Humanity has certainly shown a flair for imbuing physical objects with unsuspected magical or supernatural powers, but can we casually dismiss their existence as flights of fancy? Certainly some of them existed, and some of them have astonishing stories to tell.

* Please visit Scott's website: *INEXPLICATA-The Journal of Hispanic Ufology* - inexplicata.blogspot.com

Angel's Hair Fall

In Nebraska, USA, on November 8, 1961, a pale, fibrous substance fell on the farm of Theodore Goff at Chadron, Nebraska. The stuff fell all over the farm, draping over machinery, etc. The Air Force was notified but the investigative team which was supposed to visit Goff never showed up. Two weeks after the incident, some of the fiber was still on various pieces of machinery and Goff was trying to preserve them, hoping to get some light shed on the mystery by the Air Force.

In an article in the *"Chadron Record"* for 30 November, mention was made of the fibrous metallic-appearing stuff having been found on the Sandhills Golf course around November 8—also, Goff talked to duck hunters who said they had seen similar material near Sand hills. An unconfirmed report from the state of Georgia said fibers appeared there about 28 November.

We will reiterate at this point our request for samples of this strange material. Possibly the best way to preserve it would be to use a stick (preferably cool as warmth seems to disintegrate the fiber) to gather the stuff, then put it in a sterile jar and package it to keep the light from it. Headquarters will appreciate any samples which will be analyzed with all due haste, the results being published in this periodical.

The fiber found on Goff's farm came from a "rough-appearing, ball-shaped metallic flying object" which was tumbling through the sky.

Goff said "It was going faster than any aircraft I ever saw and it just disappeared from sight when it got out of the sunlight. It made absolutely no noise. Big chunks of it broke off."

Goff said it looked "As big around as a tractor tire, four feet in diameter," and came out of the southwest, going northeast.

The object first struck Goff as being an aircraft, then as a balloon, but there was no wind and the object was traveling at a high rate of speed. The fibers were most strange – much smaller in diameter than a spider's web fiber, they were, however, very strong. They could not be broken, but a lighted cigarette severed them. They did not burn, just seemed to shrivel and split.

APRO Bulletin - January, 1962

6.

ANGEL HAIR FOUND IN VENEZUELA
Paul Dale Roberts, HPI's Esoteric Detective

During my vacation to Venezuela, I wanted to know why Venezuela was having so many blackouts. I heard rumors that the blackouts were being caused by UFO activity.

When I got to Venezuela, I talked with fisherman Mauricio Acosta. Mauricio tells me that Venezuela has been getting some unusual UFO activity. There was a triangle-shaped UFO hovering in the night sky over Fort Tiuna Military Complex. This UFO shot down what appeared to be a scanning light. It hovered over the complex for 15 minutes and shortly afterwards, there was a power outage.

Mauricio is very fascinated with UFOs. Mauricio says in 1972, when he was living in Colonia Tovar, he witnessed a golden globe UFO hovering near his home. He went out to investigate and noticed what looked like spider webbing falling from the UFO.

The UFO was seen by some other locals and for a period of 10 minutes the UFO stayed in one spot dropping off the spider webbing material. The UFO shot straight up in the blue clear sky.

Mauricio pulled out a hanky from his pocket and started gathering some of the white spider web material. As soon as he picked it up and placed the material in his hanky, the mysterious material evaporated. The material vanished right before his eyes.

Mauricio tells me that this is his only personal UFO experience. Mauricio also tells me about a motorist in the state of Trujillo that saw a disc-shaped UFO and, after the sighting, a blackout occurred in the towns nearby. They even discovered a crop circle.

I explained to Mauricio that the spider web substance was most likely "angel hair." Angel hair is usually deposited by UFOs and is known to dissolve quickly.

SOME HISTORY ON ANGEL HAIR

Angel hair has an unusual history that is associated with UFO sightings and even sightings of the Virgin Mary. Angel hair is described as having a cobweb type of texture. Some angel hair has been known to have a jelly-like substance to it.

Angel hair received its name because it is similar to fine hair or spider webs. Some UFOlogists believe that angel hair is formed when ionized air sleets off an electromagnetic field that engulfs the UFO.

The angel hair phenomenon was witnessed in Oloron, France, in 1952. The angel hair looked like large flakes and fell from a cloudless sky. No UFO was observed during this angel hair rain. Then, on October 27, 1954, two gentlemen named Gennaro Lucetti and Pietro Lastrucci were standing on a balcony at St. Mark's Square in Venice and saw what they described as shining spindles in the sky. As these shining spindles were flying across the sky, they were dropping and depositing angel hair.

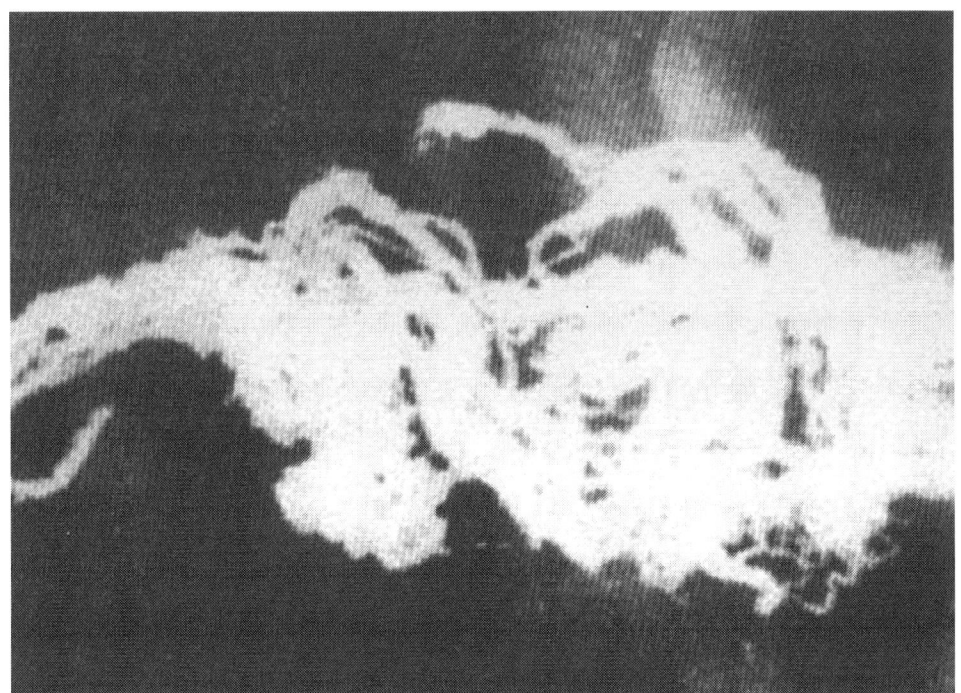

A sample of "Angel Hair" collected by journalist Giorgio Batini after UFOs were seen over Florence, Italy, on October 27, 1954.

When my wife Deanna Jaxine Stinson experienced her UFO encounter in Auburn, CA, the UFO deposited a spider web type of substance on the ground. The substance evaporated quickly. Deanna was a little girl at the time and she described the UFO as being disc-shaped and looking like a giant eye.

Angel hair is a mysterious substance and perhaps, one day, we will have the means to preserve a specimen and have it analyzed properly to determine exactly what the substance is.

Instances of this angel hair falling from the sky have been reported around the world. But studying the material has proven difficult. As the Metro points out, "The samples often evaporate before they can be properly analyzed."

However, material was collected from an angel hair incident in Italy in 1954. Spectrographic analysis revealed that the sample contained boron, silicon, calcium and magnesium. But, aside from its composition, researchers were unable to determine what it was.

Some believe that angel hair is nothing more than spider webs spun by migrating spiders that use these webs as sails to fly through the sky. But as researchers who looked at the analysis done on the Italy samples point out, spider webs are made up of nitrogen, calcium, hydrogen, and oxygen–not the elements reportedly found in the collected samples.

Halo Paranormal Investigations

www.cryptic916.com/

Sacramento Paranormal Help

www.facebook.com/HaloParanormalInvestigations/

Email: jazmaonline@gmail.com

HPI Schedule: jazmaonline.boards.net/thread/1748/hpi-schedule-2021

Sacramento Paranormal Haunted Hotline: 916 203 7503

7.

ANGEL HAIR FALLING FROM THE HEAVENS OR UFOs?

By Alejandro Rojas

France in the fifties was overwhelmed with UFO reports. One of the lesser known and more mysteries anomalies connected to this wave was the collection of white fibrous material that fell from the heavens. Dubbed Angel Hair because it falls from the sky, it was seen in conjunction with UFO sightings, but more often without. Even the Condon Report, a UFO study done by the University of Colorado in the late sixties at the behest of the U.S. Air Force, examined the issue. They described angel hair as "fibrous material which falls in large quantities, but is unstable and disintegrates and vanishes soon after falling."

One of the earlier more fantastic reports of angel hair took place in Oloron, France in October of 1952. According to several witnesses they saw a large cylinder in the sky at a 45 degree angle. Below it was a cloud and witnesses could also make out a mass of smaller objects. Using opera glasses they could see that these smaller objects were red spheres surrounded by a yellow ring. Their movements were described as "following a broken path characterized in general by rapid and short zigzags. When two saucers drew away from one another, a whitish streak, like an electric arc, was produced between them." Witnesses estimated there were about 30 of these objects. Falling from these objects was a white hair-like substance. When picked up and rolled into a ball it turned into gelatin and then disappeared. This entire scenario repeated itself 10 days later in Gaillac, France.

Since these early recollections, angel hair events have been reported sporadically. Brian Boldman wrote an extensive article on the angel hair phenomenon for the *"International UFO Reporter"* in their fall 2001 issue. His research demonstrated that during a famous UFO wave in 1973, there was also an increase in angel hair reports. The peak of the UFO wave was on October 18, on which date there were also five angel hair reports. One of the cases was from Hamilton, Illinois. Witnesses reported seeing two large oval/oblong gray objects. The second one appeared to be covered in cobwebs. Fifteen minutes after the sighting, witnesses found cotton-like material that "became a small ball which melted as it was touched."

The Condon Report's analysis of the angel hair covered incidents from 1952 through 1955, and referred to a report which suggested that the majority of the phenomena were caused by spiders. Some spiders create webs that they use to glide through the air. A medical doctor in France made the same assumptions about the French cases, although he acknowledged that he could not explain the UFO sightings. The Condon Report also conceded, "In other cases, the composition or origin of the 'angel hair' is uncertain."

The Condon Report included analysis on angel hair samples which their team had received. They found these samples to be "space grass," which they described as, "aluminum 'chaff' of the various sizes and types used by military aircraft to confuse tracking radar..." Their report ends on this note.

Boldman's report was not able to explain the phenomenon away so easily. He noted that in several cases from Australia to Italy to Argentina, samples were analyzed and found to be made up of boron, silicon, magnesium and calcium. He also found cases in which angel hair was found to be radioactive. In one case in February of 1995 in Horsehead, New York, it was assumed that the radioactive angel hair was cotton debris from a nuclear test three days prior in Nevada.

Boldman concluded that there really is a mystery here, and that the correlation between the UFO waves and angel hair reports could not be ignored. The most recent report of angel hair like substance comes from Arizona. A witness reported to MUFON on October 10, 2010, that "for three days in a row a yarn-type material fell from the sky, but was only visible hanging off of objects when the sun was low in the sky."

This appears to be another phenomenon that can be listed as one that is unresolved and thus far the answers that have been found only lead to more questions. The best we can hope is that witnesses that find this material quickly get a hold of researchers as quickly as possible. If Boldman is correct, the final results could be remarkable.

ALIEN ARTIFACTS

CASE HISTORIES FROM AROUND THE WORLD

Here are some cases of angel hair falls, both historical and recent, that still remain unexplained.

In 1477, in Japan, white cotton-like material fell for 6 hours after a luminous object crossed the sky

In 1596, in Japan, a great earthquake struck the Kyoto area at night and strange white hair fell over the region.

In 1702, once again in Japan, at high noon the sun changed color to a blood-like red and strings of a substance similar to white cotton fell to the ground.

In 1945, in the US, a man was hunting when he saw a UFO land in a clearing in the woods. The craft then emitted a humming sound, began revolving and ascended vertically. As it disappeared it discharged a shower of silvery thread-like material.

In 1968 in Canada, a farmer saw three football-shaped objects. Two of the UFOs appeared to be connected by a "long, white arc or loop" which appeared to fraying, with the third object separate. Afterwards, long white filaments fell upon the farm.

In 1971, in Australia, silvery-white globes were reported. Many appeared to be "double" with a joining thread or cord, moving around each other. The objects were seen moving in separate directions, and also changing direction suddenly (which seems to argue against the wind as the propellant). Pieces of "fairy floss" were found on the ground, which melted when touched.

At 2:00PM, on October 22, 1973, in Sudbury, Massachusetts, a child ran into the house calling to his mother to come outside to see "the biggest spider web in the world." The mother discovered in her yard a silvery-white web-like material covering bushes and hanging from the trees. As she looked toward the sky, she witnessed a shiny, silvery, spherical object moving off to the west as more of this web-like substance fell from the sky for another two hours. The witness took samples on construction paper and placed them in a glass jar and into the refrigerator, then taking them to a local laboratory for examination. The material was white and translucent and diminished rapidly. (UFO Investigator, March 1974)

At 5:04 PM, on Sunday, August 9, 1998, according to *USA Today*, dozens of residents of Quirindi, Australia, called Australia's National UFO Hotline to report that they saw cobwebs fall from the sky after twenty silver balls passed overhead.

According to the Tamworth, N.S.W. *"North Daily Leader,"* "Mrs. E. Stansfield said that she saw cobwebs falling from the sky after the silver balls passed overhead. Her daughter, who was outside at the time, was covered in fine strands of a cobweb-like material. When she tried to pick it up, it disintegrated in her hand. The family car was also covered in the strange substance."

In 2000, residents of two northern Italian towns reported an unusually loud boom, followed by a shower of "long sheer white filaments drifting down from the sky."

The Day UFOs Dropped Angel Hair Over Tuscany
By Richard Padula

In 1954, a football match ground to a halt when unidentified flying objects were spotted above a stadium in Florence. Did aliens come to earth? If not, what were they?

It was 27 October 1954, a typically crisp autumn day in Tuscany. The mighty Fiorentina club was playing against its local rival Pistoiese.

Ten-thousand fans were watching in the concrete bowl of the Stadio Artemi Franchi. But just after half-time the stadium fell eerily silent – then a roar went up from the crowd. The spectators were no longer watching the match, but were looking up at the sky, fingers pointing. The players stopped playing, the ball rolled to a stand-still.

One of the footballers on the pitch was Ardico Magnini - he was something of a legend at the club and had played for Italy at the 1954 World Cup.

"I remember everything from A to Z," he says. "It was something that looked like an egg that was moving slowly, slowly, slowly. Everyone was looking up and also there was some glitter coming down from the sky, silver glitter.

"We were astonished we had never seen anything like it before. We were absolutely shocked."

Play was suspended because spectators saw something in the sky, according to the referee's match report.

Among the crowd was Gigi Boni, a lifelong Fiorentina fan. "I remember clearly seeing this incredible sight," he says.

His description of multiple objects differs slightly from Magnini's.

"They were moving very fast and then they just stopped. It all lasted a couple of minutes. I would like to describe them as being like Cuban cigars. They just reminded me of Cuban cigars, in the way they looked."

The incident at the stadium cannot simply be interpreted as mass hysteria - there were numerous UFO sightings in many towns across Tuscany that day and over the days that followed. According to some eyewitness accounts a ray of white light was seen in the sky coming from Prato, north of Florence.

A sketch of UFOs over the stadium by Silvio Neri.

Another man who relishes the chance to speak about that day is Roberto Pinotti, the president of Italy's National UFO Centre. He has written many books about UFOs and his home in the centre of Florence is stuffed full of alien memorabilia, posters of old Italian B-movies, framed newspaper articles and black-and-white photographs of blurry flying saucers.

"The players and the public were stunned seeing these objects above the stadium," Pinotti says.

"At the time the newspapers spoke of aliens from Mars. Of course now we know that is not so - but we may conclude that it was an intelligent phenomenon, a technological phenomenon and a phenomenon that cannot be linked with anything we know on Earth."

He's also intrigued by the material that fell from the sky - what Magnini describes as silver glitter.

"It is a fact that at the same time the UFOs were seen over Florence there was a strange, sticky substance falling from above. In English we call this 'angel hair'," says Pinotti.

"The only problem is after a short period of time it disintegrates." As a 10-year-old-boy he witnessed this phenomenon himself. "I remember, in broad daylight, seeing the roofs of the houses in Florence covered in this white substance for one hour and, like snow, it just evaporated.

"No-one knows what this strange substance has to do with UFOs."

Variously described by witnesses as similar to cotton wool or cobwebs, the substance was hard to collect because it disintegrated on contact - but some people were determined to find out what it was.

One of them was a journalist at the Florentine newspaper "*La Nazione*," the late Giorgio Batini. In 2003 he told an Italian television programme, Voyager, how on that day he received hundreds of phone calls about the sightings. From the offices of La Nazione in the centre of town his own view of the sky was blocked by the Cathedral, so he went up to the top of the newspaper's building to see what everyone was talking about. The 81-year-old recalled seeing "shiny balls" moving fast towards the dome of the Cathedral.

Batini ventured out to investigate. He came across a wood outside the city that was covered in the white fluff. He gathered several samples by rolling them up on a matchstick, and took them to the Institute of Chemical Analysis at the University of Florence. When he got there he found that others had done the same.

The lab, led by respected scientist Prof Giovanni Canneri, subjected the material to spectrographic analysis and concluded that it contained the elements boron, silicon, calcium and magnesium, and that it was not radioactive. Unfortunately this did not provide any conclusive answers - and the material was destroyed in the process.

So it all remains a mystery. No matter what the scientists say, those who were there are convinced that what they saw was unlike anything on earth.

Source: *BBC World Service Sport* - Published, 24 October 2014 - www.bbc.com/news/magazine-29342407

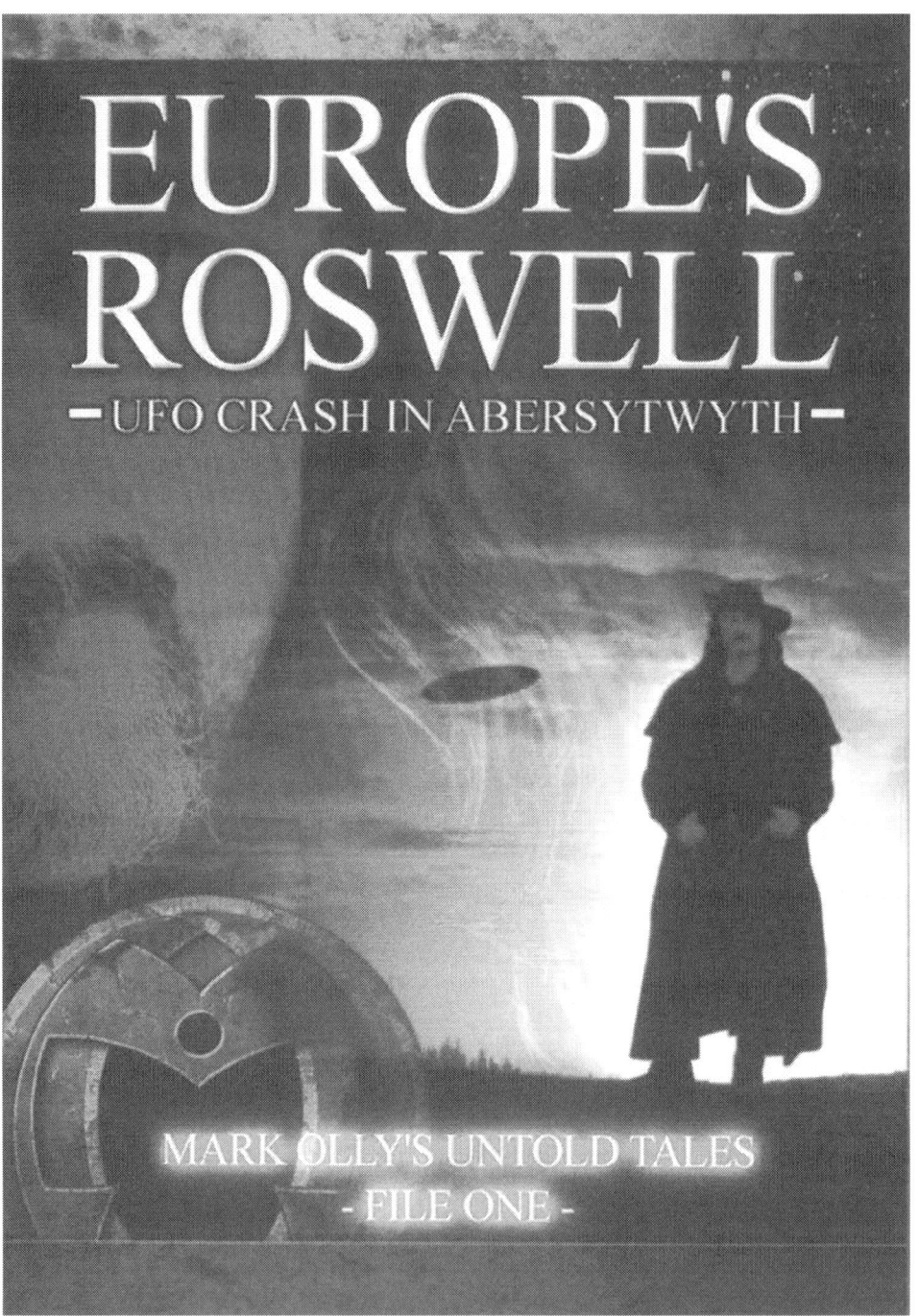

**DVD cover for Mark Olly's "*Europe's Roswell -
UFO Crash in Abersytwyth.*"**

8.

EUROPE'S ROSWELL – UFO Crash In Aberystwyth (Revisited)

Original Reports By Mark Olly & Gary Rowe Up-dated 2022.

Orbes Volantes Exstare

(Flying Saucers Are Real)

It was a bleak autumn day in 2005 when the phone in my office rang and it was Gary M Rowe, founder and director of 'The Forward To Aquarius Group' paranormal and psychic research organisation asking had I given any more thought to investigating an Unidentified Flying Object crash in the sleepy village of Llanilar just inland from the coastal holiday resort of Aberystwyth in mid Wales?

As an archaeologist I had already decided to take a small team down to the location at some point and try to recover any material left at the site – the thought of actually digging up a UFO was just too tempting to pass up – but the more we researched the less likely that this appeared possible. I just remember Gary saying *"I've still got some bits of the UFO"* and from that moment on I was hooked.

The few facts that emerged from our conversation that day seemed to point to something hitting trees during the first week of January 1983 scattering debris over four fields and a wooded copse, and then simply flying off unaffected. All we had was one newspaper report, one independent eye-witness to the scene, and one set of fragments recovered by Gary and the rest of his team from the Welsh Federation Of Independent Ufologists (WFIU). But that's not the whole story.

It all began when one of the world news-clipping agencies that operated at that time sent Gary an article from the "*Sunday Express*" national UK newspaper released on the 23rd January 1983.

Although the date and attribution was hand-written on the photo-copy we later established that the author, Andrew Chapman, had genuinely written the piece for that edition, but he could find no record in his note books for the original source of the story. He guessed that it had just been passed to him to write up.

Just like the famous Roswell incident in 1947 this press report was there one day and gone the next, and no other newspaper appears to have run with the story. "*The Sunday Express*" article is therefore crucial evidence and reads thus:

"***Strange debris out of the sky.*** *AN ASTONISHING sight greeted farmer Irwel Evans as he trudged across his fields to tend his newly-born lambs. Hundreds of pieces of honeycombed metal foil were strewn over an area the size of three football pitches. Huge twisted alloy plates, painted green on one side grey on the* other, lay every-where. And in a nearby copse branches had been sheared off trees."

"Mr. Evans phoned the police. Soon his farm at Llanilar, near Aberystwyth, Wales, was like a set from a spy thriller. Police took away fragments of metal for analysis. A team of uniformed RAF men with plain clothes officers combed the land and nearby woods using flashlights as darkness began to fall.

"**Baffled.** Among the pile of debris taken away was an aerial and a large chunk of metal with part of a serial number on it. Everyone concerned was convinced that whatever it was that covered Mr. Evans's field had fallen out of the sky at dead of night. But after two weeks the riddle still remains. Police are baffled. So, too, are the RAF. No-one in the close-knit Welsh community heard a plane that night. Nothing unusual showed up on RAF radar scanners.

"Mr. Evans, 29, who farms his 260 acres single-handed, said: 'Whatever tumbled from the sky broke up on impact. It must have been a fair size. Wreckage was scattered across four fields. Had it hit a building there's no doubt the devastation could have been terrific. It must have come down the night before I found it for the area was clear in the afternoon when I checked the flock. Yet I heard nothing at all unusual. Although the pieces themselves were extremely light they must have fallen with some force to sever branches off trees. It is all very disturbing.'

"Mr. Emyr Hughes, secretary of the Cardiganshire farmers' union, said: 'I've asked the Ministry of

Defence for an explanation, but so far have had no reply. The RAF say they had no aircraft out at the time this debris must have landed, nor were there any manoeuvres. Not only that, their radar scanners picked up nothing unusual.'

"Meanwhile, villagers are still speculating about the debris. Could it be part of a large weather balloon? 'No,' say Aberystwyth police. 'Too much metal.' Part of a satellite: 'Unlikely. Any remains would be charred. We have no explanation as yet. It's baffling.'

"An RAF spokesman said: 'The debris certainly had nothing to do with us. We are examining the fragments to try to piece them together in the hope of a clue to where it came from and what it is.'"

Still in January of 1983, and within days of receiving the article, Gary talked to Irwel Evans on the phone, then Gary and his hastily assembled crash team made their way to the site.

Irwel told them that the incident must have happened in the dead of night as there was no sign of the debris late on the previous evening. Fearing a plane crash he had quickly searched for the main body of the aircraft but failed to find it, there were no engines or mechanical components of any kind anywhere. None of the debris had rivets or resembled anything he had ever seen before.

The clean-up operation had taken a full day and involved teams of police, uniformed RAF men, and plain clothed individuals who seemed to be giving the orders. Irwel described it to Gary as "*... a scene straight from a James Bond movie.*"

Whatever the flying object was, it had clearly collided with the trees, cutting an obvious avenue of damage through the woods.

Irwel then led the team across the four fields describing the plates as *"... green on one side and coated with some hard grey substance on the other."* He had found huge quantities of metallic foil, some neatly honeycombed between two layers of the hard grey substance. The pieces looked like shattered glass with jagged edges and some of the shattered and twisted plates were over six feet in size! The overall impression he got was that some large aircraft must have exploded above the area.

The farm was bordered on the South West side by a mixed wood copse owned by the Forestry Commission and, as the team approached the field boundary by the trees, it became apparent that whatever the flying object was it had clearly collided with the trees. In an approximately twenty-five foot wide swath the flying craft had cut a straight line running through the wood causing an obvious avenue of considerable damage, some trees were knocked down, uprooted, and lying in a direction pointing towards the fields. Further into the copse the tops of many trees had been sheared off, bark had been stripped from thinner trees, and broken branches and twigs were everywhere right up to edge of the fields.

After an hour of searching it became obvious to Gary and his team that the fields had been extremely thoroughly cleared of debris but the woods promised to be another matter having been mostly cleared using floodlights as winter darkness fell. And sure enough the team managed to recover at least six pieces, two of honeycomb foil and four of shattered plate.

Just as they had been told, all the pieces had fragmented edges and one honeycomb foil section was as light as a feather but so strong that it could not be crushed.

A small section of thin metal plate only two and a half inches across and the thickness of 100 gram paper could only be bent using great force and could be used to cut material like a craft knife.

The other thicker plates looked painted on one side with dark green paint and thinly coated with a plastic honeycomb pattern on the other. In all probability the honeycomb foil had been attached to this substance at some point.

Gary later observed that the material was all curved, concave or convex, and looked shattered even though it would clearly not shatter, so perhaps the trees had somehow punctured the craft causing a pressurized explosion which effectively "blew out" the debris.

Some of the material then went privately for testing to Gary's team contacts in the aero industry and armed forces who observed that the metal forming the sheets was a type of Duralumin (Aluminum alloy) not inconsistent with material used to build fighter aircraft, but they expressed surprise at the pure quality of the sample.

They were unable to identify the green paint or the grey refluxed surface, but were adamant that *"...this green surface had not formed part of the outside skin of any aircraft as the paint was not sufficiently aerodynamic."*

Normally the only areas of an aircraft's construction that would require honeycombed or sandwiched strengtheners would be 'control surfaces' such as tail rudders or wing flaps. One expert added that *"... there is no known aircraft that could lose that quantity of its control surfaces, so as to cover several fields with metallic debris, and survive. I suggest you go back and look for something big with engines sticking upright out of the ground."*

Gary and his team did indeed return a second time to Irwel's fields but were unsuccessful in finding more crash debris so planned a third visit only to be told by Irwel that the Forestry Commission had arrived and set about removing the wooded copse. In a phone call with Gary later that day they said it was *"...because of wind damage."* No further site searches were made.

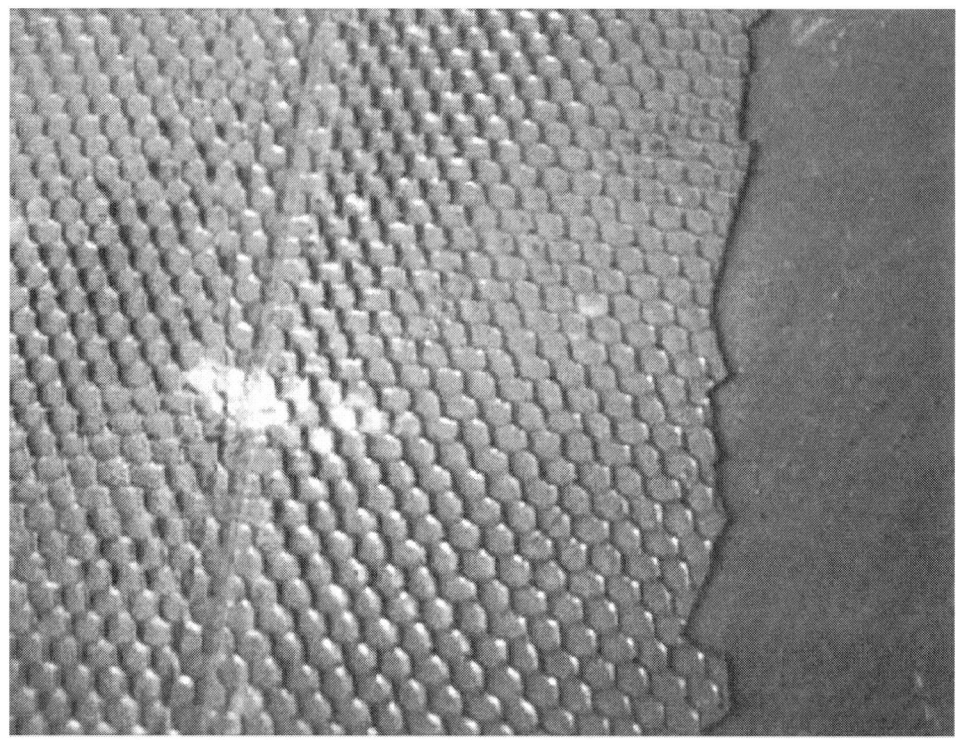

The shattered plate was taken to be tested and was found to be a type of Duralumin (Aluminum alloy) similar to material used to build fighter aircraft.

For the DVD release of *"Europe's Roswell"* my production company MythCo partnered with Reality Films to make the one hour documentary intended to be released for the 25th anniversary of the crash in 2008. We explored and re-examined every available possibility.

By this time Irwell was over 50 and really didn't have anything more to add to the statement he made to the authorities, press, and Gary at the time, indeed some of his recollections felt dulled by the passage of time. He didn't want the actual location broadcast, wasn't that interested in appearing on-camera, and neither were any of his neighbors all of whom claimed to have seen and heard nothing (as in the original newspaper report) or to be new to the village.

It was going to be a difficult production as clearly there was nothing new to add so we made the documentary following Gary's initial low-key instructions and sticking strictly to material from the time. When the DVD came out we even included all Gary's interview footage in unedited form as an 'extra' to show we had left nothing out and had not tampered with any of his statements using clever editing.

We invited professional academic researcher and Welsh based author Scott Lloyd to check out the material we had from the newspaper article and he drew a decisive blank. No one had records of anything; police, RAF, Farmer's Union, journalist Andrew Chapman, or the Sunday Express, and no other press local, regional, or national, appeared to have run the story in January 1983. Presumably a reporter local to Llanilar with connections to the *"Sunday Express,"* or even someone involved in the clean-up operation, had passed the story in virtually complete form to Andrew Chapman when the press team returned from the Christmas/New Year holiday break, but that person has never been identified.

Shortly after recovering the debris Gary had a home visit from two un-marked blacked out vehicles without plates and a knock on the door from 'the men in black' requesting he hand over the material.

Enter the key ring fragments. Gary simply pointed out that the material had already been distributed to multiple sources who would not hesitate to reveal it to the media if anything further were to happen, so the men simply got into their vehicles and drove away. He still does not keep the debris at his home address. We had the crash debris so we took the story at face value and produced the documentary anyway.

A recent 2022 search of MOD records under a Freedom Of Information request has also drawn a blank but there have been a couple of tantalizing hints that things were happening around Aberystwyth at that time.

While speaking at a UFO conference a lady came to me with her story of how, as a student studying at Aberystwyth University, she had woken up one morning in January 1983, opened her bedroom curtains, and witnessed a UFO rising out of the sea in Cardigan Bay. She grabbed a camera and photographed it but I never received the promised copy of the photograph and she never followed up on her promise to contact me after the conference.

Apparently this is not the first time UFOs have been seen in and around Cardigan Bay, and another UFO was reported further north that January over Bowring Park, Liverpool, by an elderly witness who was woken by lights at night and observed it through binoculars before it sped off across the adjacent M62 motorway. I have been told on frequent occasions that there was a period of high UFO activity over mid Wales that January but, beyond that, specific details have not been forthcoming over the years, and there was also a well attested and confirmed mass sighting of 'flying triangle' craft in South Wales on the 19th. But for the last time, this event has nothing whatsoever to do with the "Berwyn Mountain Incident!"

One expert suggested that "No known aircraft could lose that quantity of its control surfaces, so as to cover several fields with debris, and survive." It was suggested researchers return to the field and look for something big with engines sticking out of the ground.

Shortly after *"Europe's Roswell"* was released a UFO web site (that shall remain un-named) produced a written report on the crash very similar to the one contained here but at the end had a crashed disc sticking out of a forest with alien bodies hanging out of it – clearly some form of sensationalist invention or deliberate disinformation.

Another site, ufoinsight.com, is currently carrying a much more credible report with a previously unseen SUFON 2018 YouTube interview with Gary at his home talking about the incident for the last 8 minutes or so (from 58.30 minutes onwards), but the content is still essentially the same as the DVD more than ten years on. What this does clearly show is that over the years Gary has never changed any of the details of his story or the reporting of the incident.

Eventually *"Europe's Roswell"* found its way on to television in the form of pay-to-view where I believe it still remains, but when it was fully released in the UK and USA it was met by utter silence. I had just one request for an interview with a US radio station who then never followed it up. Silence.

That tells me two things: that we got the production absolutely right in every respect and there was an on-going concerted effort to cover the story up and that, at the time, no one wanted to know that UFO's were physically real and that someone actually had fragments. In the public eyes the search and mystery was very much more desirable than any physical evidence - they much preferred the *"truth to be out there – SOMEWHERE."*

I still have a small fragment of the debris, as noted Gary very sensibly distributed key rings with samples inside just in case his primary fragments were ever seized.

Over the last 15 years I have often wondered if the unidentified hexagon pattern is a primitive version of Graphine, does it conduct electricity or signals like some giant flexible computer circuit, how could the Duralumin be such a pure form back in 1983, how is this material so resistant to force but could be smashed like eggshell by low-impact tree branches, what substance is the grey-brown resin, and where did the original craft so silently go after discovering the danger of trees?

Next year in January 2023 it will be 40 years since the original crash which reminds me of the relentless passage of time – something I have come to believe is possibly the key to Unidentified Flying Objects that have plagued our world for so long with their extraordinary behavior – are they just a race of past or future humanoids who have conquered the problems of time and space? Only time itself will tell. All I can say is that, based on the physical evidence, I think flying saucers are most probably real.

MARK OLLY – SPRING 2022.

PRODUCER/DIRECTOR '*EUROPE'S ROSWELL*'

* Mark Olly is a writer, TV presenter, public speaker, lecturer, and archaeologist with over 30 years' experience working in history, media, and the arts. His films inclued: "The Life & Times of the Real Robyn Hoode" (2015) and "Europe's Roswell: UFO Crash at Aberystwyth" (2009).

The Amarillo (Texas) *Sunday News-Globe* of April 9, 1950 Carried a Headline: "So You Saw a Flying Disc? –This Boy TOUCHED One!"

By Gordon Tompkins, Jr.

Two young boys from River Road went fishing late yesterday morning, but instead of a string of perch, they came home with a flying saucer story.

This flying saucer landed. One of the boys touched it.

The pair are David Lightfoot, 12 years old, son of Mr. and Mrs. J. A. Lightfoot of Bluebonnet Drive and Charles Light foot, 9, son of Mr. and Mrs. o. W. Light foot of River Drive. The boys attend River Road School...David in the sixth grade, and Charles in the third. They are cousins.

The two excitedly babbled out their eye-witness of the saucer to newsmen and radiomen yesterday afternoon. The accounts, except for minor differences in measurement estimates, were alike.

In substance, this is a composite of what David and Charles said happened to them yesterday morning: The boys went fishing shortly before 11:00PM yesterday morning on a creek near the southern boundary of the Convalescent Home northeast of the city.

Before they had pulled in even a little one, they sighted what they thought at first was a balloon. It was about 20 degrees above the horizon, they indicated by arm motions. The object was traveling from the south. As it came nearer, the object decreased in speed, and it became apparent that it wasn't a balloon.

The disk passed by only a few feet over the boys, and David shouted to Charles, "I'll bet that's one of those flying saucers, Charles, I'm going after it." Charles did not follow. David said the disk circled slowly and disappeared over a small hill to the north. Before David could top the rise, saucer had landed.

"It was about as big around as a regular automobile tire," the youth remarked and about as high as my knee – maybe a foot and a half. I could see it good."

According to David's account, the object was rounded on the bottom and had a top part which resembled a flat plate. The top, he said, was separate from the bottom by a space perhaps one-half inch in depth, and was held to the bottom section by "some sort of screw or something in the middle."

"When I first saw it good," David explained, "the bottom was still, but the top was spinning around real fast, and on the top of the part that was spinning around there was a little peak that had a kind of spindle sticking out of it. The spindle was still, too. It must have been connected to the bottom part."

The object was blue-gray in color, and had no openings of any sort other than the space between the top and bottom section, according to the boys' story.

"When I came over that hill, the thing was only a little ways from me and I ran a ways and dived at it," David related.

"My fingers just barely touched it and it felt slick, sorta like I guess a snake would...it was hot, too."

The youngster went on to relate that before he could "get a hold of the thing, the top began revolving faster, and it made a sort of whistling noise and took off without warming up or anything."

It disappeared in a straight line into the northeast in a matter of five to ten seconds, he estimated.

In the process of taking off, though, the object emitted some sort of gas or spray that turned the boy's arms bright red and caused small welts on his arms and face. David's father confirmed that detail.

"When the boys came running home, talking about this flying saucer, one of the first things I noticed was that his arms and face were red, and there were welts on them."

David explained that after his father had applied a skin balm on his face and arms, the welts gradually disappeared, but the red remained.

Both David and Charles were sincere in telling the story. However, neither of the boys believes the alleged flying saucers are from some other planet.

"I think it's something the United States is doing," David said.

APRO Bulletin - January, 1963

9.

MINIATURE UFO CAPTURED IN JAPAN
The Little Known Kera UFO Incident
By Tim R. Swartz

In late 2019, Tom DeLonge's *To The Stars Academy* signed a contract with the U.S. Army to collaborate in the study of "exotic metals" – i.e., parts of crashed UFOs. Both parties hope that this agreement will lead to the development of advanced technologies. However, a group of Japanese schoolboys in 1972 did what To the Stars Academy and the military have not yet been able to do (at least to the best of my knowledge): they actually captured a miniature UFO.

But they lost it.

Only to recapture it.

And then they lost it again.

Several times.

Missed it by that much…

This case is little-known in the West and has been largely forgotten in Japan. Nevertheless, this incident is one of most bizarre and frustrating cases in modern UFO history.

ONCE UPON A TIME ON SHIKOKU ISLAND

Kōchi City is the capital of Kōchi Prefecture on the Shikoku island of Japan. With its subtropical climate and facing the Pacific Ocean, the area is well-known for its seafood and as a popular gathering place for surfers from all over Japan.

Nearby is the Kera area, with a total population of about 20,000. Kera is a quiet, residential area, considered to be one of the best places to live for those who work in the neighboring cities.

This suburb became the focus of attention starting on August 25, 1972 when 13-year-old Michio Seo, who was on his way home from school, spotted a strange metallic object hovering in the air over a rice field that ran along the side of the road.

Seo watched as the strange, hat-shaped device flew back and forth over the overgrown field. Seo later compared the movements of the object to that of a bat making sharp turns in pursuit of insects.

Intrigued by what he was seeing, Seo tried to get closer to the small craft, only to be repelled when it shot a dazzling beam of light at him. Seo knew that the bright light was a warning for him to keep his distance, so he quickly headed home and told his friends about what had just happened.

Along with his friends -- Hiroshi Mori, Yasuo Fujimoto, Katsuoka Kojima and a friend named Yuji, Seo returned to the field around 7:00 PM in hopes of seeing the tiny UFO again. At first the object was nowhere to be seen, but the group decided to wait. After about an hour, the craft returned and, to their amazement, it flew back and forth over the rice paddy, pulsating with a bright, rainbow colored light.

The boys were able to take several photographs of the tiny UFO as it hovered a few feet over the rice field.

When one of the boys got brave enough to try and approach it, the small disc began to emit a loud "popping" noise and glow a bright blue. This sudden change was unsettling enough that the frightened boys decided to call it a night and quickly run to their homes.

However, their curiosity got the best of them and the children returned to the spot over the next several days, hoping to once again catch a glimpse of their mysterious visitor. On September 4 the craft returned, but the boys were again too nervous to approach it. They vowed the next time to bring a camera and prove that what they were seeing was real. On September 6 their vigilance paid off when they found the craft sitting once again in the field.

One of the boys pointed their camera and took a photo. When the flashbulb fired, the UFO began to spin and quickly rose into the air. Another photograph was taken, but this time, when the flashbulb went off, the small object fell to the ground and continued spinning for a short while in the soft earth.

Fourteen-year-old Hiroshi Mori cautiously approached the now dormant object and picked it up. Later, he said that it felt like something was "moving" inside. One of the other boys took a photo of him holding the craft.

The boys took turns holding and examining their prize. Finally, Miro wrapped it in a plastic bag and took it home for a closer examination. It was determined that the silver-colored object weighed about three pounds, was around four-inches across and eight-inches high. The underside had circular grooves...almost like a vinyl record. In the middle was a square with 31 holes surrounded by three unknown symbols.

CAPTURED AND RECAPTURED

At this point, the boys felt that it was important to reveal their find to their parents, who had grown concerned about the boy's unexplained nightly activities. Yasuo Fujimoto's father, Mutsuo, was the director of the Center for Science

Education in Kochi. They figured that he would be the best candidate to try and identify the object.

"The frequent nights out of the boys began to worry us parents," Mutsuo Fujimoto said. "I told my son, if it was true what he said, to bring the object to me. He did, and it was something like an ashtray, cast iron, but too light for this metal. It was impossible to open, and inside were pieces similar to a radio. I did not give it more importance, but now I regret not having studied it more closely."

After it fell to the ground, 14 year old Hiroshi Mori picked up the tiny UFO. He later said he felt something "moving" inside.

After Mr. Fujimoto's examination, the craft was again placed into Mori's backpack...only to mysteriously vanish the next day.

Over the next few weeks, the boys would spot the little UFO, or one that looked like the first one, flying over the rice paddy. They noticed that it never appeared on rainy days, which led them to come up with a strategy in hopes of recapturing their prize.

On September 19, they returned to the field, carrying a bucket of water and some rags. Once again they discovered the device sitting quietly on the ground. Quickly covering the object with the rags, the boys poured water over it, and then turned it over and poured water into the holes underneath. When they did this, the tiny UFO began to brightly glow and a loud buzzing sound burst from the interior. The group retreated, fearing for their safety, and began throwing rocks at it until it quieted down.

Satisfied that it was now inoperative, they took it to Katsuoka Kojima's home for a closer look.

FULL OF ELECTRONICS

Examining their recaptured UFO, they attempted to look into the tiny holes underneath. Inside they saw what appeared to be levers, symbols, and electronics. Next, they forced a wire into one of the holes and hung it upside down. This caused the top of the dome to slightly separate from the lower section. The boys could see "complicated electronic equipment" as well as an unidentified viscous material. They feared that this was the remains of a tiny alien pilot they had accidentally melted when they poured water on it.

Then, with a large hammer, they struck the object several times to test the exterior. Despite the apparent low density of the metal, the hammer bounced off with no damage or even marks left on the craft.

To further test the UFO, the boys wanted to place it in the oven and then the refrigerator, only to be stopped by Kojima's mother, who refused to allow such experiments using her appliances.

Since there was to be no further testing that night, the boys wrapped the disc up and covered it in pillows, thinking that this would help prevent any exposure to "atomic radiation" that could be coming off it. The object was given to Seo and Mori for safekeeping, while the rest returned to their homes for dinner and homework.

When the group returned later that night and checked under the rags and pillows, they were shocked to discover that the disc had once again vanished. At this point they realized that the tiny UFO could be some sort of robotic device that was being operated by someone, or something, using remote control.

HIDE AND SEEK

Over the next week the miniature UFO seemed to be playing a game with the boys. It would appear in places where they frequently played, allow itself to be captured, only to disappear from bags, cupboards or whatever method the boys thought up to try and keep it safe. During this time, the friends decided that they should mark the silver dome with paint to confirm that they were actually finding the same UFO again and again. The boys had lost and found the object so many times at this point they naturally assumed that if it disappeared they would eventually find it again near the paddy field or in one of their backyards.

It seemed, though, that whatever was controlling the UFO was growing tired of the game. The group had taken to sealing the object in plastic and then placing it inside a bag of water. They would then make sure that the bag was physically tied to the wrist of one of the boys.

All of these precautions seemed to be of little avail, as, on the evening of   September 22, the team met for a bike ride in the city of Kochi. They decided that all of them would take turns carrying the device, which they no longer left unattended.

On close examination, the boys said they could see "complicated electronic equipment" within it.

The knotted bag containing the UFO was put into a canvas bag and placed into the bicycle basket of the carrier. The bag went from cyclist to cyclist as they cruised through the city, heading for a local bike shop. Suddenly, the boy holding the bag, so to speak, said that he felt that the rope attached to his wrist was being forcefully pulled.

The boys immediately untied the rope and opened the bag, but when they looked inside, the little UFO had once again mysteriously vanished. This time, however, the disappearance was permanent...the boys would never see their intriguing little disc again.

ALMOST FORGOTTEN

It's not known whether or not any UFO groups in Japan investigated this case afterwards. Word certainly got out, as there is part of a documentary available on YouTube that was obviously filmed some years after the incident. In the film, the boys, who appear to be high school age or even older, recreate the events. Their parents and other witnesses are also interviewed. Unfortunately, the YouTube version has a narration that covers the original audio track and it is impossible to tell what the witnesses are saying.

In 2004, "*UFO Comics*" in Japan published an illustrated account of the case. This introduced the encounter to a whole new generation of UFO enthusiasts. In 2007, Shinichiro Namiki – the director of the Japan Space Phenomena Society (JSPS) – reopened the investigation. The head of the JSPS Osaka chapter, Kazuo Hayashi, was sent to speak with the remaining witnesses. All maintained that everything had happened as they had originally reported.

BATTERIES NOT INCLUDED

The Kera UFO case is certainly unique. Obviously, some skeptics denounce it all as simply a hoax, perpetrated by a bunch of bored schoolboys. It does sound suspiciously like some of the plots of popular Japanese science fiction television shows like "Ultraman" and "Kamen Rider." If it wasn't for the parents, who later confirmed having seen and handled the disc themselves, it would be very easy to dismiss everything as a prank.

There have been other cases involving miniature UFOs, and it can be presumed that these tiny UFOs are some kind of remote controlled (or even autonomous) "drones," as their size would seem to prevent having any sort of living pilot inside. That being said, I have uncovered some unbelievable cases of mini-UFOs along with their tiny pilots. Even more amazing these incidents also have school kids trying to actually grab the tiny humanoids and the little ships that they rode in on.

Ahmad Jamaludin, who wrote the 1981 "*A Summary of Unidentified Flying Objects and Related Events in Malaysia* (1950-1980)," published an article in the November 1979 issue of *MUFON UFO Journal* titled "*Humanoid Encounters in Malaysia.*"

This piece has reports exclusively of small UFOs and mini-humanoids, and when I say mini, I mean really mini. All are reported to be six inches or smaller. Here are some of the cases that Jamaludin uncovered.

CASE 1. Johore Bahru, 1970

Four boys going to school one morning saw a small UFO and tiny 6-inch tall creatures. The boys quickly reported it to the headmaster and, since it occurred in the school premises, the news broke out and the whole school was combing the area looking for the creatures. One small burnt patch on the ground was found. The UFO was gone.

CASE 2: Gambang near Kuantan, 1973

Two schoolboys claimed to have seen tiny humanoids only 3-inches tall in the school compound. It is said that one of the creatures was actually caught by the boys, which attracted the attention of a teacher. He arrived in time to see it before the tiny creature finally managed to escape.

CASE 3: Bukit Mertajam, 1973

A group of boys playing football in a school field sighted a small UFO that landed nearby. From it emerged tiny creatures, which one of the boys immediately tried to catch. A beam of light was fired at the boy's hand. The UFO flew away.

CASE 5: Miri, Sarawak, 1973

Several boys saw a humanoid about 6 inches tall wearing a white suit and cutting a fence wire with an intense beam of light. They tried to catch it but it was lost in the bushes. No UFO was sighted.

CASE 6: Miri, Sarawak, 1973

Several people on vacation along the beach sighted a group of tiny beings wearing white suits. There were about six or seven of them. They wore no mask and resembled humans except for their tiny size. This group consisted of possibly males and females. The females were notable because of the long hair. An attempt to catch them failed. No saucer was sighted.

CASE 7: Bukit Mertajam, 1979

A small UFO landed on a field near some boys. A boy tried to catch a 3-inch creature that emerged from it. It fired a beam that temporarily paralyzed his right arm. The UFO then flew away.

There were not less than seven reported landings in Malaysia from the first known landing in 1970 through May 1979. From these reports there were five cases of close encounter with tiny humanoids and two other cases from East Malaysia where no UFO was sighted.

The creatures measured, in all cases, either three inches or six inches tall. All were equipped with a type of ray gun. They were described as well dressed in one-piece suits. Some had slightly larger heads and round eyes. The 3-inch humanoids sometimes had two antenna-like structures protruding from the head. Cases 1,2,3,4 and 7 occurred in West Malaysia, all reported by schoolchildren, and all landings took place in school premises.

As well, Jamaludin included some additional cases in an article for *Fortean Times* (Number 35, 1981) called *"Tales from Malaysia."*

1. Kulim, May 1979

A group of boys playing near a cattle-shed sighted a small object, less than two feet in diameter, gliding over the fields and emitting a strange sound. The UFO had three legs of its landing gear extended, seemingly about to land, when one of the boys tried to touch it.

An intense beam of light shot out from the rim of the craft and temporarily blinded the boys. They screamed for help, attracting the attention of two adults, who came rushing to the scene and also observed the UFO.

Meanwhile the cattle in the adjacent shed were intensely restless and were struggling to break away from the posts to which they were tied.

2. Bukit Mertajam, May 29, 1979

On 19 May, at 3 PM, six schoolchildren encountered a landed UFO and four three-inch-tall entities near it. One of the students, it was reported, tried to catch one of the creatures but was shot in the hand.

The creature then fired another shot at a brick, breaking it in two. After the shooting, the creatures scurried back into the tiny object.

Another pupil grabbed the UFO with both hands but had to let it go when he felt what seemed like an electric current passing through his hands.

The object then took off, leaving in its wake a shower of falling leaves. A 21-year-old youth playing nearby came to the scene after hearing the commotion just in time to see the shooting and the object taking off.

With these strange incidents from Malaysia in mind, remember that the boys involved with the Kera UFO said that they saw a "viscous" substance inside the disc...which left them worried that they may have accidently melted a tiny extraterrestrial pilot.

Considering that the UFO phenomena has all sorts of mind-bending facets that leave researchers in a permanent state of puzzlement, it is not that outrageous to consider that at least some of the mini-UFOs that have been spotted over the years might be piloted by tiny creatures whose origins are unknown.

So the next time you are taking a stroll outside, walk lightly, as you might accidently crush a tiny interplanetary (or interdimensional) craft that has landed to take in the sights of Planet Earth.

Even though you would then have your very own "Alien Artifact," cleaning your shoes off afterwards would certainly not be an enviable task.

* Portions of this chapter were originally published in *"Tim R. Swartz's Big Book of Incredible Alien Encounters"* — 2019, Global Communications/Inner Light Publications

Approximately six-months after the Kera, Japan, UFO encounter, a similar miniature UFO was photographed at Suonenjoki, Finland.

10.

CAN WE BELIEVE THE ACCOUNTS OF CRASHED
UFOS? – FROM ROSWELL TO CENTRAL PARK
By Sean Casteel

In a book from Global Communications called *"The Case For UFO Crashes: From Urban Legend To Reality,"* author and publisher Timothy Green Beckley argues that the many stories and rumors about the UFO crash phenomenon are basically true and that the government is hiding its knowledge of that truth because of National Security concerns. And just how does Beckley make his case?

By a combination of both anecdotal evidence – the stories that are told about the numerous crash incidents – and with evidence of a more concrete kind – actual government documents that have been leaked or obtained through Freedom of Information Act requests.

THE STORIES PEOPLE TELL

At this point, it may be relevant to say something in defense of anecdotal evidence. Budd Hopkins, the late abduction researcher, once stated that while no one has ever provided absolute proof that UFOs and their attendant phenomena are real, nevertheless there is plenty of evidence, the kind of evidence that would convince any jury in any courtroom that

UFOs did indeed exist and represented an alien or otherworldly technology.

Hopkins went on to say that ALL evidence presented in a trial is itself anecdotal. For instance, when the prosecution presents DNA evidence against a defendant, the jury never actually "sees" the DNA evidence. What they get instead is a law enforcement DNA expert telling them a "story" about that evidence. In other words, all the jury really has to go on is the anecdotal testimony of a scientist, the spoken words of an allegedly credible spokesperson who "talks" about the evidence in an attempt to bolster the prosecution's case.

So it is with the UFO phenomenon. We often have only the stories people tell to go on, but ignoring those stories is a self-defeating way to close our eyes to some very important truths.

In *"The Case For UFO Crashes,"* convincing anecdotal evidence abounds, sometimes from sources that command our respect and are seemingly beyond reproach. Beckley writes about the late astronaut Gordon Cooper, who stated publicly that he first encountered UFOs in the 1950s while he served as a jet fighter pilot stationed in Germany. He and his comrades-in-arms witnessed an over-flight of strange objects that could stop on a dime and make 90-degree turns in the middle of their flight paths. Cooper later said that reports of crashed UFOs seemed credible to him and the entire subject warranted further study.

Beckley also quotes movie producer Peter Kares, who claimed to have known an ex-Air Force pilot who was at the scene when a crashed disc was carted away.

"The Case For UFO Crashes: From Urban Legend To Reality"
by Timothy Green Beckley

"Later, he was harassed," Kares said, "sent to a psychiatrist and nearly drummed out of the service because he refused to sign a pledge that he would never talk about what he had inadvertently seen. We were even able to talk with a full colonel who claims he saw with his own eyes a UFO that was being kept in storage. In one case we investigated, motion picture footage was taken of several UFOs traveling along at speeds upwards of 10,000 miles per hour."

When Kares followed up on what he had heard, he was told by the government that no such film existed, which the producer attributes to the government's not wishing to cause panic in the streets.

THE AZTEC CRASH AND JOHN PEELE'S DISCOVERY

Beckley also provides a fascinating retelling of the Frank Scully story. Scully was a Hollywood reporter who found out about a saucer crash said to have happened in Aztec, New Mexico, in 1948 that left several alien corpses behind. The resulting scramble by the military and FBI to cover up the incident and discredit Scully and his sources makes for an interesting case history of just what happens in the wake of a UFO crash. (To learn more about Scully and to read the book he wrote about the Aztec crash, read the Global Communications bestseller *"Behind The Flying Saucers*, Updated Edition.")

In chapter after chapter of *"The Case For UFO Crashes*," Beckley recounts the details of his own investigations into the phenomenon: the many leads he followed up on, the extensive correspondence and phone calls with witnesses and others privy to information on the subject, as well as the dead ends when important sources would apparently die or simply disappear along with whatever proof they had hidden away from their relentless pursuers. Beckley's tireless, dogged pursuit of the truth produced no dramatic "smoking gun," but it does serve as evidence that something is going on, that the tip of an iceberg of information is undeniably there.

One of the more fascinating stories told in *"The Case For UFO Crashes"* came to Beckley when he was in Florida doing media appearances to promote his now defunct magazine, "*UFO Review*." One of his regular correspondents told Beckley about a health food store owner in Orlando, Florida, who made an important discovery in 1977. While jogging across the United States to promote exercise and fitness, John Peele discovered what he thought was a UFO crash site near the California desert town of Octotillo, a short distance from the Mexican border.

"There, jutting up out of the desert," Peele told Beckley, "were several large pieces of what resembled Plexiglas."

Peele had served in Vietnam as an army helicopter pilot and knew immediately that the Plexiglas was not from any earthly military aircraft. He also discovered pieces of lightweight metal similar to aluminum that were honeycombed on one side and refused to bend as such a light metal should.

But the real surprise was Peele's finding a glove similar to the pressurized gloves worn by our high-altitude test pilots and astronauts. This would not be too unusual except for the fact that the glove was in miniature, as if it were intended to be worn by a child.

"Of course, children do not pilot high-altitude planes," Beckley writes, "nor does the government allow individuals under a certain size to join the service, ruling out that the glove might have been manufactured for a midget."

Peele found another glove a few feet away, but that one appeared to have been badly burned in the crash. Since Peele was jogging at the time, it was impossible for him to cart the material away, so he buried it with the intention of returning later in his vehicle to retrieve it. However, when he attempted to go and reclaim the artifacts, a peculiar storm came up, rendering the desert skies pitch black, a phenomenon unknown to even the locals.

More details of Peele's incredible story can be found in chapter 11 of this book.

THE MYSTERIES OF HANGAR 18

As further crash lore, there is of course the well-known belief that the government conceals a great deal of evidence in Hangar 18 at Wright-Patterson Air Force Base in Ohio, or alternately, in a location on the base called "The Blue Room." The late Arizona senator, Barry Goldwater, openly declared that, "I have never gained access to the so-called 'Blue Room' at Wright-Patterson, so I have no idea what is in it. I have no idea of who controls the flow of need-to-know information because, frankly, I was told in such an emphatic way that it was none of my business that I've never tried to make it my business since."

One wonders if Goldwater's diplomatic use of the phrase "emphatic way" is a euphemism for some kind of violent threat. Just what lengths do the secret-keepers go to in order to maintain their hold on UFO information? Who, if any, among our elected officials are ever told what the military is concealing in terms of crashed UFOs, alien bodies, or anything else to do with the phenomenon? Former president Bill Clinton stated on at least two occasions that he was denied access to classified information both on the Roswell incident and alien-related activity at Area 51 in Nevada.

GOVERNMENT SECRECY IN WRITING

The book also contains additional anecdotal evidence, stories that sound like science fiction but have been handed down through various sources as true. There is the night a UFO came crashing down over an Ohio shopping mall, for example, or the unbelievable eyewitness account of a UFO that fell inside New York City's bustling Central Park after being shot at by the military. One story details the rescue of a living alien from a downed spaceship as it rested on a military runway in New Jersey. The UFO pilot later died in captivity, but not before witnesses saw what happened and eventually went forward with the story. Allegedly there exist photos of an entity named "Tomato Man," another crashed UFOnaut, photos that to this day have never been satisfactorily explained.

As promised, *"The Case For UFO Crashes"* also contains a prodigious amount of government documents from the files of the Department of Defense, the FBI, and even a bizarrely named agency called "The Interplanetary Phenomenon Unit." The reader can see actual files written by – and correspondence between – members of the military and various intelligence agents as they struggle with responding to the alien presence while at the same time keeping the truth from the very citizenry they are sworn to uphold and serve.

The reprinted documents, including the legendary Majestic-12 material, are numerous and too many to go into detail about here. The book also includes an interview with Ryan Wood, a researcher, investigator and author who has made a career, alongside his father Robert, of vetting and authenticating leaked government documents on UFOs as well as others obtained through FOIA requests, but their main emphasis is on the MJ-12 papers.

The younger Wood's interest in UFOs began at age 15, when his father brought well-known UFO researcher Stanton Friedman home for dinner. At the time, Friedman worked for Robert Wood at the McDonnell-Douglas aeronautics company researching antigravity. As an adult, Wood says he now specializes in analyzing documents while ignoring other aspects of UFOlogy.

"We don't do abductions," he said. "We don't do lights in the sky. We don't do anything other than military and intelligence history as it relates to the Majestic documents. That's our focus."

In the interview, Wood tells the story of how the MJ-12 documents were initially obtained when Navy and Marines veteran Timothy Cooper began to receive portions of them in his mail; Cooper has since amassed the largest collection of MJ-12 documents and original Blue Book files in the United States. It is not clear why Cooper was chosen as a conduit of the leaked files, another mystery to add to the already large pile. As of when the interview with Wood was conducted, the flow of documents to Cooper continued, giving the Woods and their team plenty to keep them busy.

Woods offered one particular document as an example.

"One document that is very interesting," he said, "is called The Interplanetary Phenomenon Unit Document, and it's an intelligence summary. It's a draft, an assessment. It has fourteen points. It was written on 22 July 1947, after the team went to the Roswell area to deal with some of the crashes."

The document is very specific, detailing the longitude and latitude of Mack Brazel's ranch, where the Roswell debris was strewn over a wide area, and pinpoints the location of a second crash near Socorro, New Mexico, around that same time. Meanwhile, Wood is confident that the saucer crash documents are not "disinformation" or "psychological warfare," saying he doesn't think that over the long haul it would be to our government's advantage to use fake documents simply to lie to us, our allies or even our enemies. Somehow, backstage and under a heavy covering of secrecy and classification, a real life drama is being acted out that supersedes any attempt to deceive those outside the loop.

WHY ARE THERE CRASHES AT ALL?

Given that UFO crashes happen on a worldwide basis, resulting in many headaches for governments both allied and "hostile" to the United States, the question must be asked, are the aliens simply inept pilots of their own spacecraft?

"That's another good question," Wood responded. "Why do they crash? I'm speculating. The data shows that there are crashes. I think it has to do with the human anthropomorphic view of the alien agenda. We think they don't want to crash. We value human life. We rescue our pilots from the ocean or from behind enemy lines. These alien bodies are disposable, biological robots that are like toilet paper to us. Their mission in the universe is to hop from galaxy to galaxy, star system to star system, gather information and move on.

"And they may be less well-equipped than with a mother-ship for interstellar space. They may be using something more like a scout ship, and when they encounter radar or some freak lightning bolt, they malfunction and have a problem.

"Then of course there's the deliberate thought – that they're crashing on purpose, because it helps man develop new technologies and advances our civilization without shocking us. It may take decades or centuries to understand and catch up, but it's like you're shown the future. Then you begin to do the reverse science thing, or the reverse engineering."

It's hard to imagine the UFO crashes are intentional and have the benign purpose Wood is talking about, but one quickly learns when studying UFOs that nearly anything is possible.

COMING FORWARD WITH THE "TRUTH"

Wood feels some of the documents may be leaked by people who regard it as a matter of conscience to do their part in revealing the UFO phenomenon to the world.

"People feel that it's wrong to hide the fact that we're not alone," he reasoned, "to have hidden the greatest technological advances from the masses and in essence slow the advance of global humanity by scores of years.

"What they did," he continued, "which is hide it all, forced a few cloistered scientists working in secret to push the ball down the field with a paperclip instead of hitting it with a baseball bat. That's really the great crime, that they've hidden the technology and the evidence and thwarted the standards of civilization."

In the interview with Wood, he also speculates on the assassination of JFK and the mysterious death of Marilyn Monroe as being connected to MJ-12. Were the two murdered to prevent them from going public with what they knew about UFOs and the government cover-up?

Such questions continue to haunt us regardless of the fact that they may seem ridiculous on the surface, even crazy or in the category of "fringe beliefs." Still, what Timothy Green Beckley has achieved in *"The Case For UFO Crashes: From Urban Legend To Reality"* deserves an open-minded look. There are any number of reasons that flying saucers come hurtling from the sky, and just as many reasons for the truth of that to be withheld from us.

RECOMMENDED READING

THE CASE FOR UFO CRASHES – FROM URBAN LEGEND TO REALITY

BEHIND THE FLYING SAUCERS: THE TRUTH ABOUT THE AZTEC UFO CRASH, EXPANDED EDITION

Could the strange silver gloves found by John Peele actually be evidence of extraterrestrial visitors? Or are they proof of a secret space program active in the 1960s? (Photo is a reproduction)

11.

THE ALIENS GO HAND-IN-GLOVE WITH MYSTERY
By Sean Casteel

Before he died in 2021, Timothy Green Beckley had been in the UFO game for 60 years and it was hard to impress him with anyone claiming to have the "real thing," or concrete proof that aliens had visited Earth and left behind artifacts that irrefutably demonstrated that reality.

BY WAY OF INTRODUCTION

In an interview Beckley conducted, along with his co-host Tim R. Swartz, on their podcast "Exploring the Bizarre," one of the rarest examples of just such an artifact was the subject of their conversation with John Peele, a health food restaurateur and long distance runner, whose story was spine-tingling in the extreme.

"It's a very strange and odd case," Beckley began. "One of the most peculiar I've been involved in. It includes meeting Mr. John Peele in late 1978 or early 1979 in Florida. At the time, he owned a health food store and restaurant and had a very unusual artifact that he had located while running across the Arizona desert. It is one of the few alien artifacts that we can actually point to as existing.

"We hear about the Roswell UFO crash," Beckley continued, "and deceased alien bodies strewn across the desert, but there's very little proof – Roswell fanatics will go for the jugular on this one – there is very little proof that any of that took place."

Beckley was editing *"UFO Review,"* a tabloid newspaper focused on the titular phenomenon, and had published a small number of books, when he first met Peele while on a promotional tour and fact-finding mission in Florida. Along the way, Beckley got Peele's phone number and Peele invited him to visit his home.

BECKLEY SEES THE ARTIFACT

"He showed me this – well, I call it an alien artifact – I don't know what else you would call it – and it took my breath away," Beckley recalled. "There's been nothing like this before."

Beckley said he has discussed the find with several people and has been greeted on occasion by total disbelief. Where is the artifact? Where is the witness? Decades later, Beckley found Peele on Facebook and invited him to appear on "Exploring the Bizarre."

Peele found the artifact in 1977, while on a long-distance run from Daytona Beach, Florida, to Santa Monica, California. (Peele is such a well-known cross country runner that he was consulted by the makers of the hit movie "Forrest Gump," who wanted firsthand information on what it's like to run the kind of distances portrayed in the movie by actor Tom Hanks.) Peele and his running companion averaged 48 miles a day.

In Arizona, Peele and his companion parted ways, which they often did, having run out of conversation after spending so much time side-by-side.

"You go your different ways and discuss it later," Peele explained.

PEELE STUMBLES INTO RESTRICTED TERRITORY

Peele said he had unknowingly run into a government restricted area, where he came upon railroad tracks that resembled a roller coaster ride.

"It was pulled out," he said, "and because of the way that railroads are connected, one rail to the next, it literally made loops. It was quite bizarre and I did not know what I was seeing at the time. I'm still not a hundred percent sure. But I went over towards that and started following it. And at one point I came upon a large – what looked to me, and I'm a fighter pilot, so I say this with some knowledge – I saw what looked like a wing."

The "wing" was extremely thin and, though it was eight-feet long, it was so light that Peele was able to use two fingers and from the end hold it horizontally with no problems. The wing was also very strong but it was still possible to bend it. He began to gather other little "remnants" of what he felt was an obvious crash. There was a kind of Plexiglas or carbon glass present that was unbreakable, although some of it had shattered from impact into the rocks in the desert.

"But you could pick up a piece of it and you could not break it," he said. "I didn't bring any of that. I was a little bit concerned that this might be radioactive. I didn't know what I was dealing with."

AND THERE IS THE GLOVE

"And then I found the glove," Peele said.

There were actually two gloves. Again, not knowing if the objects were radioactive, and whether he might have to at some point simply get rid of the gloves, Peele hid one in a small cavern under some rocks. He acknowledges that the markings he used wouldn't be there now, but he has enough of a visual memory stored in his mind that he feels confident he could find the glove and wreckage debris again.

One of the two gloves was in much better condition than the other, so, in case he eventually had to get rid of them, he preferred to keep the one in good condition hidden away in the Arizona desert. Since it turned out he was able to keep the glove, he said, he should have brought the good one home.

In any case, the gloves were made with some sort of a hide. They had a pressure valve on the outside/backside of the hand. There was a zipper that worked from the inside and enclosed something very similar to Nomex, a fire retardant fabric first developed by DuPont in the 1960s.

The glove was so small that there was no way it could have fit a human hand. There was a prehensile or "opposable" thumb, meaning a thumb that can be placed opposite the fingers of the same hand, thus allowing the digits to grasp and handle objects. Opposable thumbs are characteristic of primates, such as human beings, and were a crucial step in our evolution.

"When you hold it," Peele said, "you understand that it is not from Planet Earth. You can't look at it and see how it was made and see how detailed everything is. But there's no hand that would fit it that could be used for any particular reason that we know."

The glove also had what seemed like writing, much of it incomprehensible and most likely nonhuman words. But Peele was able to read the English words "secure" and "large."

A 'CHILLING' FIND

At this point in the interview, Beckley interjected, "In this field, physical evidence like an alien artifact is almost impossible to come by and is always in dispute. But there is something so strange and peculiar about this that it almost sends chills up and down one's spine. OK, here's the glove. We don't know who it belonged to. Would an alien race be that close to us that they would have a glove? And have four fingers and a prehensile thumb? It would indicate to me that we might be talking about something from the future or something from a parallel universe."

Peele replied, "I've certainly given the parallel universe concept much thought. First of all, there were two gloves. One was almost perfect. The other one had gotten a little – obviously from the crash – it had not really burned but it had been heated and altered. They were basically the same size, so neither one had shrunk."

After meditating from the age of eleven on, Peele said he was very familiar with the energy that is a crucial part of meditation.

"You hold that glove and you can feel it," he said. "Even after all it had been through, when you held that glove it had like an energy of its own."

WHERE IS THE GLOVE TODAY?

Beckley asked if Peele still had the glove.

"Unfortunately," Peele answered, "I believe that the glove – I had a divorce. As divorces aren't, it wasn't very good. And a lot of things disappeared, and I'm pretty sure the glove went with the ex. She says not, but I'm pretty sure that's where it is. I have wanted to go back. I have so many friends who say, 'Let's go out there. Let's get the other one.'"

"Do you know where it is?" Beckley asked. "Don't tell us the nearest town because there would be a hundred people out there tomorrow. Did you ever do any research on the railroad trestle or whatever it was?"

Peele said that those were the days before the Internet and there was no way to do a search for information like that. However, he remains certain that he can find the glove with no difficulty.

Beckley asked Peele if he believed the crash debris was from an object from outer space.

"The glass was probably a metal-impregnated plastic," Peele answered. "But you couldn't break it. It was blue, which we don't use in our aircraft windows. The wing portion that I found, like I said, it was probably somewhere around six to seven to maybe eight feet long. I don't remember exactly now because I never had a measuring device with me, but I know it was taller than me."

Peele tried an experiment with the wing material. He put it between two rocks and sat on it. It bent a little, he said, but it didn't collapse. Again, he could hold it aloft horizontally with only two fingers in spite of its size.

"Our wings are made with titanium," he said, "and we have all sorts of materials that make them strong but light. But nothing like that."

MORE DETAILS ON THE GLOVE

Returning to the subject of the glove itself, Peele described it in more detail.

"It felt like leather," he said, "and looked like it. It was smooth on the inside, kind of rough and grainy on the outside. So the smooth side was in, and the bladder that had the pressure capability went into all the fingers and the thumb and across the back of the palm. Inside that was a string that would pull and zip it. It was definitely an internal zipper and it was pretty interesting.

"Today you'll find these little zippers that zip across and have no teeth. They just seal. It was almost that sort of thing, though it was much stronger. But it had no teeth. You could just seal it. You could pull it and it would close it.

"I had people from several of the space development teams look at that. And everybody was impressed. Nobody could decide where it came from and they all said it did not appear to have been made by anybody under government contract. And the word 'Large'? If that was large, you can only imagine what small would have been."

THE GLOVE IS SEEN BY OTHERS

Peele was not subjected to a frightening visit from the Men-In-Black, nor did he feel he had been approached by anyone using false credentials. Everyone he loaned the glove to eventually brought it back, though it frequently took much longer than he was told to expect. When the various agencies did return the glove, it was always because they had more questions they wanted answered. But he was never able to obtain anything in writing from the various groups investigating his find.

To all who encountered it, the glove indeed seemed to have a life of its own.

"That's what was so cool about this glove," Peele said. "People would hold it and say, 'Ooh, it's like my hands are tingling.' Everybody who saw it – I don't want to say they were freaked out, but they WERE freaked out.

"A large number of people did see it," he continued, "because I would show it in my store. It was called 21st Century Foods and was the first vegetarian restaurant and health food store in the area. I would keep it in the restaurant there and people would come in. They had heard about it because it did have some news media coverage."

People would ask to see the glove and it sometimes caused some of the curiosity seekers to become so unnerved that they would exit the restaurant.

"Most people watch 'Star Trek' or whatever," Peele explained, "but they don't really want that to be real. Other people DO want it to be real. I've always felt we cannot be egotistical enough to think we are the only humanoid-type beings in the universe. Think about the variety of people on our planet and then expand that to over a few thousand light years away. We have to have others that are similar to us – more advanced, less advanced. Not as nice, nicer. They're out there."

A STRANGE KIND OF 'DESERT STORM'

Shortly after discovering the crash site and the two gloves, Peele made a return trip to the area, accompanied by his former wife, who was driving a motor home to provide mobile shelter to Peele and his running companion. Peele said he knew exactly where the site was.

"I was going to zip over, get some of the components, and bring them back," he said. "Looking up into the sky, there were definitely some things that would not be normal under any circumstance. My former wife saw a lot of that as well. I was running to the site. I got really close to it. I saw these 'things' in the sky. It was really bizarre."

The winds suddenly came up with unusual strength. It didn't actually rain, but there was water in the air.

"It was like swirling water," Peele recalled. "It was really weird. I turned around and went back because it didn't seem like a safe environment to continue in. My ex was so freaked out when I got back, looking at what was in the sky.

"The closer I got to where the glove and the other components were, the more intense the weather got. Your mind's on about 4000 things and it's easier to sit back 40 years later and think about it than it is to experience it at the moment. But there was no doubt in my mind that there was a connection

"It wasn't frightening by any stretch. I don't really have fear. I've flown fighters since I was 19, and nothing freaks me out. I don't have that, thank God. But, at the same time, I'm very aware of energy, having started meditation when I was eleven and dealing with energy and that sort of thing. So when the energy would start to build, it was an intriguing experience, but the winds got up to a point where I did not feel comfortable being there."

Peele said he did not often talk about the Arizona desert glove experience.

"But I have a certain group of people that ask me questions all the time," he said. "The reality is very simple. There are others around us and some of us are aware that they're here."

To which Beckley replied, "You certainly may have the proof – if anybody does."

UFO Landing Leaves Purple Fluid Behind

At about 8:20PM on the night of August 19, 1965, Harold Butcher, 16, was operating the milking machine at his father's (William Butcher) dairy farm, located near Cherry Creek, New York.

The boy was listening to a newscast from a portable radio, when he noted static-like interference which drowned out the program. The tractor to which the milking machine was connected then stopped. Outside the barn, a bull which was chained to an iron stake began to bellow and attempt to pull loose.

At this juncture the boy ran to the window of the barn and saw a large elliptical-shaped object with a reddish glow or vapor underneath it, as it appeared to land about a quarter of a mile from the barn. He heard a steady bee-beep sound.

The object was on the ground for only a very few seconds before it shot straight up into the air disappeing into the clouds.

Butcher, using the phone extension in the barn, notified others in the house and they came out. All noted a strange odor in the air, and the clouds into which the object had disappeared glowed a greenish color.

Approximately a half hour later, the object reappeared and seemed to be circling the area. Harold's mother called the State Police. Troopers came, and then notified the Air Force which initiated an investigation including a Captain and four technicians.

They found a purplish liquid substance in several places, small two-inch indentations in the ground, as well as patches of singed grass and shrubbery.

APRO's investigator visited the Butcher farm, found Harold to be an intelligent boy who has quit school in order to run the farm, as his father is physically unwell. The bull which initially attracted Harold's attention to the outside, was fastened to the iron stake by a rope or chain which ran through a ring in his nose. He pulled so hard that he bent the quarter inch stake over to almost a 45 degree angle.

Upon further questioning about the object, Harold said it looked "like two turkey platters" lip to lip. The purple, viscous substance amounted to about a quart. Harold said that when he put one of his finger into it, it was as if someone was putting pressure on the finger.

The Butcher boy also told our investigator, Perry C. Euchner, that the 16 cows, which normally yield from three to four cans of milk per day, were yielding only approximately one and a half cans per day for almost a week after the incident.

Source: *APRO Bulletin* - November/December 1965

12.

JOHN LENNON'S EGG
By Nigel Watson

"There's UFO's over New York and I ain't too surprised

Nobody told me there'd be days like these

Strange days indeed -- most peculiar, mama"

"Nobody Told Me," featured on John Lennon's *"Milk and Honey"* album (1984)

Many decades after the break-up of The Beatles, their music continues to have an important impact upon our society and culture. In the 1960s they spoke for a whole generation who wanted "peace and love," and they reflected a general belief that revolutionary changes were afoot.

Obsessive fans have certainly regarded The Beatles – and particularly John Lennon – as the messiahs of a New Age, and have searched for meanings and messages in their body of work. At the most trivial level, their records have been endlessly played backwards and forwards in order to discover any hidden messages. Their album covers have also been analyzed and discussed to root out similar nuggets of secret information. Even at the start of Beatlemania, the Rev. David Noebel told his Baptist congregation in Claremont, California, that The Beatles were using a synchronized beat to put American youth under the influence of mass hypnosis. He announced that "These Beatles are completely anti-Christ. They are preparing our teenagers for riot and ultimate revolution against our Christian Republic."

The science fiction writer Philip K. Dick received more benevolent enlightenment from the songs of The Beatles. At the beginning of 1974 he received mystical messages from a force he called the Vast Active Living Intelligence System (VALIS). He believed that thousands of years ago a craft sent by three-eyed crab-like beings from the star Fomalhaut came here and helped to create our civilization. They returned home leaving VALIS to inspire and cajole key individuals to keep us on track. The communications from VALIS often came to him at night and provided him with information about the aliens' complex control of humanity. One such message came to him whilst he was listening to the song "*Strawberry Fields Forever*" on the radio. VALIS used The Beatles song to tell him: "Your son has an undiagnosed right inguinal hernia. The hydrocele has burst, and it has descended into the scrotal sac. He requires immediate attention, or will soon die." He rushed his son to hospital and found that this diagnosis was true. VALIS, through The Beatles, saved his son's life.

Intriguingly, Philip K. Dick's VALIS visions and mystic messages eventually faded away six years later, which mirrors the time span from when Lennon saw his UFO over New York to the year of his untimely death.

The writings of the ex-wife of David Bowie, Angela Bowie, asserts that David Bowie, Galileo, Isaac Newton, Bob Dylan, Jimi Hendrix, Ghandi, Winston Churchill, and John Lennon were inhabited by alien presences. These aliens, whom she calls The Light People, look upon our human activities and intervene to help us evolve and progress. To this end they seek out people at critical moments in history and light them from within.

UFO ENCOUNTERS

We are on firmer ground with Lennon's most famous UFO sighting. He publicly announced his experience on the rear cover of his 1974 *"Walls and Bridges"* album: "On 23 August 1974, I saw a UFO J.L."

At the time he was estranged from Yoko and living at 434 East 52nd Street, New York, with May Pang. They had spent the day at the Record Plant East recording studio working on the *"Walls and Bridges"* album. It was a hot evening and at about 8.30PM May went to take a shower whilst Lennon looked through artwork for the new album. Not long afterwards he had an urge to look out of the window that led onto their apartment's roof terrace. In the meantime, May had just got out of the shower when she heard him call out for her. She quickly found him naked on the roof terrace looking at a strange object in the sky. It was not unusual for Lennon to go on the terrace in this state of undress as it was protected from public view, but the object he was pointing to in the southeast sky was a different matter.

She said it looked like a large flattened cone with a bright red light on top of it at a distance of only 100 feet away. The object came towards them and as it did so it revealed a dazzling row of flashing white lights along its rim. Lennon thought it was so close you could almost touch it and Pang confirmed that it was so close you could have thrown something at it. The craft, estimated to be the size of a

Learjet, moved silently throughout the sighting. Lennon reported that, "It was flying slowly, about 30mph, below, I repeat, below most rooftops (i.e. higher than the 'old' building, lower than the 'new') all the time it was there, I never took my eyes off it."

They tried taking pictures of it with a Polaroid camera but this was broken. Nevertheless, they managed to take a few shots with a 35mm camera. After a few minutes the UFO passed the United Nations building, veered left over the East River and, after moving over Brooklyn, it merged with the air traffic over southern Long Island.

John Lennon and his girlfriend, May Pang, claimed to have seen a UFO over New York City on August 23, 1974.

A bit later that evening it returned. On the second occasion they were ready for it and had a telescope set up to watch it. Unfortunately, the light from it was so bright that no further details could be seen through the telescope than by the naked eye.

They were sure it was not an aircraft, balloon or a helicopter. Evidence for the sighting has proved elusive. Lennon rang his friend, rock photographer Bob Gruen, urgently asking him to come round to develop some film. When he developed these pictures they came out blank. This is just as well, as they would have depicted a naked John Lennon with his arms outstretched shouting at a UFO "Wait for me! Wait for me!"

On behalf of Lennon, who wanted to remain anonymous at this time, Gruen rang the local police to see if anyone else had seen a UFO in the area that night. They said they had received two other reports. On ringing the *"Daily News"* they said that five people had reported seeing a UFO on the East Side. But when Gruen rang the *"New York Times"* they hung-up on him when he mentioned flying saucers.

The sighting made him even more fanatical about the subject. A few weeks afterwards he made an audiotape of his thoughts about UFOs. On this he states that the UFO he saw was part of a bigger fleet that took their energy from the nearby nuclear power station at Indian Head. He concluded that they are from another planet and that they are covered-up by governments because "they would threaten the status quo" and would cause a revolution on a political and personal level.

ALIEN ARTIFACTS

When ufologist Larry Warren interviewed May Pang in 1988 he was surprised to learn that this had not been Lennon's first alien encounter. On several occasions Lennon said that he suspected that aliens had abducted him during his childhood because he always acted and felt different from other people. Unfortunately, he never went into any detail about his suspicions.

ALIEN EGG

The famous spoon-bender and UFO contactee, Uri Geller, claims that the rocket pioneer Werhner Von Braun showed him a piece of a crashed spaceship at the Goddard Space Flight Center. Taking it from a safe located in a basement of the building, he asked Geller to see what impression he got from touching the object. Geller confirmed; "I felt it wasn't terrestrial. It was metallic, elongated and had a hue I have never seen before." It seemed like it was breathing and alive. He added: "The surface had a pearl-like quality that almost seemed to be in 3-dimensional color."

Although that artifact is allegedly hidden away, Geller does own John Lennon's alien egg. The story is that as they ate at a restaurant in New York, on an unspecified date in early 1975, Lennon confessed to Geller that he believed in life on other planets. Taking him into his confidence, Lennon said he had a weird encounter with some alien beings in his Dakota Building apartment. One minute he was fast asleep with Yoko Ono, the next he was wide-awake.

A powerful searchlight seemed to be shining outside the bedroom door. Light poured through the keyhole and around the edges of the door, and at first he thought it was either a searchlight being used by intruders or a fire in the apartment. On opening the door he found four thin, human-sized bug-like entities. They had small mouths and big bug-like eyes. He described them as scuttling around like roaches. He went to get them out of his apartment but before he could even touch them some form of invisible power held him back. They gently held his hands and pushed his legs into a tunnel of light where he was shown an "outstandingly beautiful" movie of his life. After that his mind went blank and the next thing he knew he was lying on top of the bed covers next to Yoko.

Lennon was sure he had not been tripping or dreaming. Indeed, he was adamant that he had never seen any creatures like this before in his life. The story was so bizarre that he had only confided in two people before he confessed to Geller. When he told Yoko about it she believed him but the unnamed other person (May Pang perhaps?) didn't believe him, thinking he had been "high."

As evidence of this encounter he found an egg-like metallic object in his hands when he woke up next to Yoko. The solid object with no visible markings was too weird even for Lennon! He speculated that it was a ticket to an alien planet, but he was quite happy with New York. He gave the alien egg to Geller, who has kept it ever since. Geller notes that "When I hold the cold, metal egg in my fist, I have a strong sensation that John knew more about this object than he told me. Maybe it didn't come with an instruction manual, but I think John knew what it was for."

BELIEVERS AND SKEPTICS

We mainly have Geller's word for his "egg" encounter, and even May Pang, who was one of Lennon's closest associates at that time, is skeptical of this story. After watching a TV interview with Uri Geller during which he spoke about Lennon's alien egg, a Danish viewer, Dann Simonsen, realized he owned a similar object. It was purchased from a souvenir shop in the late 1960s. These super-ellipsoid eggs were created by Danish scientist/artist Piet Hein (1905-1996).

Simonsen thinks that Lennon probably obtained the egg when he and Yoko visited Norden Fjord World University, Skyum Bjerge, Denmark, during the Christmas holidays in 1969. Lennon was not averse to surreal streaks of humor and could well have played upon Geller's gullibility in this matter. You would have thought that even a believer in the story would have got the object tested or examined to see if it showed any signs of extraterrestrial origin, but Geller explained that he has never had it tested because "I want to leave the mysticism around it."

Later, the arch skeptic James Randi also asserted that the egg was a Piet Hein creation. He adds that Geller's account of this story has varied but Randi provides no detail. This became evident when Geller put the egg on display in London to mark the 25th anniversary of Lennon's death. Geller reported that: "John said he had been lying in bed, when a bright sphere of light appeared in one corner of his room. An alien hand came out of it, and gave him the egg. And he gave it to me." Another, more recent version is that Lennon swore to Geller that an alien inside it tried to communicate with him.

To be fair to Geller, he might have shortened the story over the years to provide sound bites for the media. As for the origin of the object, an egg is not an unusual shape for a designer to use and the similarities noted between these objects could be a coincidence.

Michael Luckman, the author of *"Alien Rock,"* in an email to me dated 08 April 2006, notes that: "I'm urging Uri Geller to have his controversial alien egg analyzed. It would be the ultimate proof of alien visitation to Earth and would be worth millions of dollars, perhaps tens of millions of dollars if it held up to rigorous scientific testing."

Luckman also notes that Lennon's egg seems to have some type of markings on its surface, whilst the stainless steel Piet Hein eggs are entirely smooth.

When I contacted James Randi he informed me that the so-called egg is "accurately described as an ellipsoid, but I've never seen an organic egg of that shape. Thus, any attempt to relate the primal egg form to this mathematically-derived object is presumptive. Incidentally, one fact that makes this super-ellipsoid unique is that it has two axes of symmetry: It can be spun either on its side, or on its end.

"Since posting an item about this item, I've received two more sent to me from readers, one in brass and the other chrome – or nickel – plated.

"These are common souvenir-shop oddities. Either Geller is exceedingly naïve to consider it extraterrestrial, or he has chosen to be awed in order to have an attractive woo-woo story to relate. Your choice."

May Pang, John Lennon's ex-lover, is equally skeptical about the egg. She told me: "If Uri had an egg that John gave him...he would have told the world immediately and let the world see it. What was there to hide?"

Piet Hein, a Danish mathematician, poet, and designer, stumbled across the super-ellipse and decided to make it 3D. The result was the "Super-egg," an "egg" that has the ability to stand stable on either end.

The alien egg is now on display at the Uri Geller museum Old Jaffa. He sometimes allows it to be put on public display but so far he has not allowed it to be scientifically examined.

Geller steadfastly states: "I don't really care what Randi says, I believe Lennon."

AND IN THE END . . .

We can never be certain that Lennon didn't have what we now call childhood alien abduction experiences or was given an "egg" by aliens, but we do know that he saw a UFO and that the rest of The Beatles believed in such encounters. Their perceptions might have been unduly colored by the use of a range of drugs - especially LSD in the latter half of the 1960s - and the ideas of the counterculture that were swirling around at the time.

Since their rise to stardom and enduring popularity they have been increasingly regarded as virtual Gods and in some tragic instances the ultimate incarnation of evil. Their many recordings as Beatles and in their solo work is sufficient for listeners to gain evidence for their particular obsessions that often feature the Bible and fears of great and imminent apocalyptic events. This underlines the continuing potency of their work that addresses us equally on a general and a personal level. Their recordings address the feelings and aspirations of a generation that wanted to give peace a chance and a world where science fiction fantasies could come true. As yet we are as far away from global peace as we are from contacting alien beings – but these dreams continue with a little help from our friends.

* Nigel Watson has researched and investigated historical and contemporary reports of UFO sightings. He has written for numerous books, publications and websites such as Fortean Times, Strange Magazine, Paranormal Magazine and Wired. He is the author of such books as: "*Portraits of Alien Encounters*" (1990), "*Supernatural Spielberg*" (with Darren Slade, 1992), the editor/writer of "*The Scareship Mystery: A Survey of Phantom Airship Scares, 1909-1918*" (2000), and "*Alien Abductions in the USA*", published by McFarland.

Diane Tessman has experienced contacts with otherworldly beings her entire life.

13.

CHILDHOOD "SOUVENIRS" LEFT BEHIND
By Diane Tessman

Diane Tessman is a professional psychic and lifelong UFO experiencer. Here she tells the story of concrete, physical objects remaining after alien encounters in her childhood.

• • •

I prefer the term "encounters" instead of "abductions."Dr. R. Leo Sprinkle's regression notes from 1981 are toward the end. My memories under regression are at the very end.

First Encounter: As best as I can figure, at age 4, I had the first encounter. I was a little kid; I wasn't keeping track of what year it was. I was a serious, intelligent little kid. It happened in 1952 (I was 4 and possibly 5 for the second one).

In the hypnosis notes, Dr. Sprinkle asked me details, such as who else was on the ship during my first encounter, what clothing did they wear, and what not. I could see these things and I gave answers but I was less sure of these answers than I was sure about my conscious memory (it has always been with me) which is simply sitting with this being - a man - whom I describe as human but enhanced (not sure what I mean by that but I do have a theory). He was calm as he talked to me, possibly reticent (I felt). I felt sadness in him but I could have been wrong; again, I was a little kid!

He looked human except for amber eyes, not hazel but really amber. They were or seemed translucent and that mesmerized me! (I realize this is similar to Travis Walton's description of a UFO human). He was of normal build, maybe 28 or 30, maybe 5'9" or 5'10". My dad was six feet and this man was shorter than Dad.

This is the best "translation" I can make from what a small child observed. He had tawny-colored hair, casual cut hair (not super military), and he had a likeable quality.

He and I sat maybe half a foot apart on an earth-like bench. I sat on the bench, feet dangling, and I looked at a hologram-like large "frame" (the size of a human statue) across the corridor in this small earthy "terrarium" area with a plant or two. The hologram swirled in pastel colors; I don't know if it was just decorative or hypnotic or what. Calming? I waited for him to "arrive" and sit down, so there was this swirling "artwork."

He talked with me, I don't know if it was verbal words or telepathy (possibly sent in concepts, not words specifically?). He was gentle in his talk and demeanor. What he said is in the hypnosis notes which I'll enclose. Actually, he didn't say a lot that I can remember word for word but conceptually, I think a mountain-full of info was said/sent into me? (Not sure). Sitting with him, looking at him, is a conscious memory.

Note: I am sure someone might say, "Sounds like you were molested but covered it in your mind with this memory." No. Simply not. I have never been molested, never physically abused. Included with Dr. Sprinkle's regression notes is his psychological profile of me. I took tests before he would regress me.

I am not sure what exactly the "crux" of that encounter means or meant-- what "this being gave me and took from himself" (see Leo's notes), or maybe he did not actually lose anything of himself but my child-mind feared he had lost what he gave me???

I only know I cried soulfully during hypnosis and can easily bring that deep level of love - and worry for him - to the surface even now – and I felt I was home where he was, not just because of him but simply *home,* and then knowing I'd be without that home for the rest of my life as Diane – deep sobbing of a kind I don't think I've done for any other event in my life. Maybe the death of my dog Sinsee or my cat Sakima, but even that was slightly different sobbing.

Apparently more was done on that encounter (or at some point on another encounter), than him just talking to me but I have no memory of what else was done. But my unexplained missing membrane must have happened at some point. Details on that in a moment.

I had no fear or panic the whole time; I've read an article recently which says scientists can now block the fear response in humans. I have to wonder if my fear was blocked. I felt/knew there was nothing to fear. It was as if I had his understanding of Diane's situation: It would change me forever but in a good and necessary sense. The only drawback: always being different having had this encounter—having gnosis (Knowing) of life beyond present day Earth which most people do not have.

In 1979, I had a "year of awakening" to the enormity of this childhood experience. It was a year of paranormal experience, precognitive warning, and the best UFO sighting of my life. I'll send that account along with this file.

More details on this First Encounter in Dr. Sprinkle's notes at close of this file.

2nd Encounter: My family went from our home in North Iowa to the Canadian lake/woods for a two day vacation (one day and night there). We stayed in a very primitive cabin by Eagle Lake, Ontario. Canadian lakes have a history of unidentified submerged and/or flying objects. It is a very active area.

My parents, my 14-year-old brother, and I were to meet with Uncle Bill and his family at the lake. The adults didn't get along very well, not sure why we even attempted this vacation. Brother Norm was at a bad "stage" (see the Moe, Hoe and Poe addendum in a moment) and as usual, I was sort of a lone entity. My cousins, two boys, were always busy trying to kill each other.

Uncle Bill and Aunt Marge had a separate cabin with their sons. My parents left our cabin (daytime) for a while, Norm was fishing, and that left me in the cabin alone. It was unlike mom to leave me alone, she was a "helicopter mother" (hovering) so that is odd in itself. I had brought my books along in a box of family stuff. My favorite was from the Little Golden book series, "*Toys in the Attic*," which was about not fearing the unknown (in the book a little boy was afraid of stuff in the attic, a rooster statue, an Indian head-dress and other stuff).

I have never been regressed regarding this encounter. This is from conscious memory.

I was alone in the cabin and this man from my first encounter walked in the door rather quickly, almost like someone was chasing him.

He had on dark glasses, a red checkered shirt, and jeans—trying to fit in with the fishermen, methinks. I have talked with two other "experiencers" (independent of each other, two people I believe are telling the truth), who also met "future humans" and these two people also reported one being was wearing this "regulation" red checkered shirt and jeans. Is this the "uniform" for walking on Earth, 20/21stCentury? They wore sunglasses also.

I remember consciously this being/man putting the dark glasses up on his head, like people do.

I remember clearly he said, "Diane, are you frightened?"

I can see myself thru his eyes...I am holding the Little Golden book, seersucker shorts, long red braids, a little girl.

I/she said, "No." (No, not afraid).He stretched out his hand toward me.

I remembered him from "before." I felt close to *home*. I remember nothing else. As hard as I try, I can come up with nothing more. It is not just blank, there is a memory barrier.

Repeating: I have had this memory always – no regression involved. And this memory has never been hypnotically regressed.

This is a photo of Eagle Lake, showing fishing cabins on the shore.

Possible Evidence: Next I remember, I was showing my mother the "lard" that got on my favorite book. She called it "lard" actually, but we had none in our family box of snack foods, because mom was an early health food freak. She detested lard. Was there lard in the cabin? It was a primitive cabin with a big hole for a toilet (I feared I would fall in), no food provided, not even sure there was a stove. I did not put my book on a stove anyway.

I had the book until I was around age 22 when we moved to the Virgin Islands and lost stuff in the move (I am not 100% sure that's when it got lost). It was very "funny" lard. It saturated every page equally (about half the page), same design, not lessening as it dissipated. The saturation included the heavy cardboard back and front covers. It felt "soft" and was not really quite lard. I never got to examine the book in the light of my "awakening." It was lost in the interim years when I was about 22 when I took my first teaching job on St. Thomas, Virgin Islands.

Was it a byproduct of "beaming up" or ectoplasm or angel hair? (per UFO and paranormal lore).

Or just a reminder of some sort that he had been there?

The lard explanation makes no sense either and I do not think the lard was on the book yet when he first came in the cabin door.

Evidence: Ok, now the "scar." (missing membrane)

I had seen a scar on my outer mouth just above my lip, on the line of symmetry and the fold-- up to very near the beginning of my nose - since I first played around with make-up, about age 9. I asked my parents about it. I thought I had some small facial deformity at birth. It was obviously a surgical scar, not jagged. Very neat. They were doting parents. Both were college graduates, and they had no other children to be chasing.

(My half-brother was ten years older than me).

I asked them independently and together many times after I became an adult, and they both said, "No, no deformity, no procedure, and how lovely I was as a baby." Neither of them had a clue where it came from.

When I went to Dr. Sprinkle for hypnosis in 1981, I still thought it was a scar on the OUTSIDE of that area. I was curious about it then and had researched that brain surgery in 1981 was just beginning to be done by going through the back of the nose or through a nostril.

Meanwhile, I had kept asking my parents occasionally and they were genuinely puzzled. My mother disliked my preoccupation with UFOs and I think she would have gladly given me a mundane explanation if she could have to burst my bubble, so to speak.

Then about five years ago, I realized that the scar is not only on the outer skin, it is INSIDE my mouth. There is a membrane called a frenulum (a frenulum is a useless membrane) which connects gum to lip. I never missed my membrane because I didn't know it existed for most people!

I researched this membrane. It is often injured in mountain biking accidents, says Google. Then there is a procedure some mothers have done to toddlers to remove this membrane. Maybe if it is going to interfere with adult teeth? Or too tight or something? It can be snipped for a baby at birth to nurse better but mine is not snipped, it is totally gone. Plus my mother said this was not done. It apparently causes some discomfort, some blood, but is a minor procedure.

The removal is right up against the bottom of my nose. It is not a partial removal. I'm not even sure how a local 1952 dentist or doctor could get comfortably into that upper area. Why would they? I do allow that there might be an earthly explanation (perhaps done at birth) but what are the odds that my mother swore it was not done AND that I had this strange "contact" and my life was deeply affected by this long before I discovered the missing membrane?

That membrane would be ideal cloning material. The mouth is the best nourished of all the body. OR it would have to be trimmed to go thru the nose to the brain for a procedure.

OR, the membrane could have been a necessary tissue sample for re-engineering my own body's t-cells to fight leukemia. Doctors in 2021are curing cancer for many people, including childhood leukemia patients, by extracting their t-cells, re-engineering them (weaponizing them) and inserting them back into the patient to fight and kill the cancer. This is the cutting edge of our current medicine: To engineer through gene therapy, the patient's own cells and body to fight the cancer.

Future humans would have this medical technology far advanced of our own present level, but, it seems to me, they would need a tissue sample from the individual.

When I was about age 3 and 4, I had ongoing strep throat, heavy nose bleeds often, and was considered "weak." My mother even bought high topped shoes to support my ankles better. Today, when a child has sore throats, is skinny, and has frequent heavy nose bleeds, they are checked for leukemia. But I never was tested for it and my tonsils were not even taken out. They were talking of taking out my tonsils but I got suddenly "cured" or was at least suddenly over the illness.

By age five onward, I was an extremely active child, excelling at acrobats, swinging like Tarzan, and climbing trees.

OR it could simply be a reminder to "Remember! The proof is right under your nose, Diane."

I had an eye tuck in '87, the only plastic surgery I've ever had, by a respected plastic surgeon in La Jolla, CA, and I asked him about the scar (of course I still pointed to the outside of my mouth, I had not discovered the missing membrane inside).He muttered, "That is a laser scar."

1952, a laser?

Additional Facts and Thoughts: After the two encounters (I remember two, are there more?), I had invisible friends I called The Re-members. They did not really play with me like invisible playmates, they were just *there,* especially when I was outside, which was a lot.

I have the memory of a fire on the roof of our house which apparently never happened. I guess I dreamed it but it was such a sharp dream that it has caused me to have a lifelong fire phobia to the point of not wanting to light matches.

My mother had an outside door in my bedroom blocked off, which was actually a necessary fire escape route, and I've wondered if she thought someone had spirited me away for a few hours, thus my mother was protecting me from something like that happening again. And a few other odd things throughout the years, certainly many synchronicities.

I have felt a shared consciousness with this being/man all my life. I can't tell you how to drive his starship, nor the formula for free energy. However, thus happened all the "channeling" I did for over 39 years, which is considered by some to be among the best "channeling" there is. It is among the most intelligent, if I do say so myself. And my messages supported myself, my daughter and animals for countless years (I do admit to this financial fact, but I have always been sincere).

I've lived a very unconventional life. I feel the encounters changed my life. I've had non-ordinary experiences, goals and accomplishments. I have no regrets. I have forever gone against the wind, but I am full of empathy, especially for nature and her creatures. I have integrity, intelligence, and I am "different" in a way which is not a physical difference; I have always been acceptable in physical looks. I embrace the "difference" in my persona.

I do feel I have a purpose, even a mission – to raise the level of human consciousness from WITHIN the timeline, especially before it is too late for this poor old beautiful planet and her life forms.

When I think of the encounters, I often find myself looking at "little Diane" from HIS perspective. "Feet dangling off bench" is one of them. Also, in my second encounter, I plainly see Little Diane holding her Little Golden Book. This is HIS perception/memory. In general, it is not unusual for me to be looking out of what seem to be his eyes/his mind. At this point, I am him and he is me, this is natural. This is an enhancement; mental illness is about fragmentation.

There are more memories from the second encounter in my book "*The Real Life Transformation of Diane Tessman,*" available on Amazon.

Addendums: Near-Proof of UFO activity surrounding both encounters, the knowledge of such manifesting many years later:

I do want to tell you about two children who experienced the unknown, having what were almost certainly alien encounters back in 1952. One child reacted with fear and even violence, while the other child was compatible with this alien contact and wave length which was, after all, not negative.

Or at least, this is my conclusion after thinking it over for many years. The child who reacted with fear was my half-brother, who was nearly ten years older than me.

The child who seemed compatible with alien energy, finding it positive and fascinating, was me. My brother was 14 (almost 15), at the time, and I was 4 or 5 years old.

I acknowledge that my brother's experience and my experience might not be related at all, but here is my theory:

I feel that in 1951 and 1952, ETs had a program of contact with individual humans. I feel that alien contact with some humans turns out to be incompatible with the intricate working of the human mind and its electromagnetic field. The human freaks out, to put it crudely, are traumatized afterward. This "beyond frightened" bad reaction applies to some abductees, and it explains what happened with my brother. Of course, the contact could simply have been with bad aliens.

This program or coordinated alien effort, according to my theory, came to North Iowa in the years 1951 and 1952. UFO sightings and encounters do seem to happen in specific areas. I am not saying that North Iowa, and/or my brother and me, were special in anyway, but I do believe we were both contacted. We were/are both above average intelligence.

(This is the near-proof given me from "left field" nearly 60 years after my "main" encounter):

There was a saucer-landing in a hay field outside of Toeterville, Iowa, in 1952, wherein "two small midget-like humans" emerged from the saucer, scooped a sample of water from a nearby creek into a container, then walked back to their saucer and zoomed off.

This account was told to me about three years ago by an elderly man named Dick who went to high school in North Iowa with my brother. When he first stopped at my place, Dick was trying to find what had become of my brother, but then we got to talking about UFOs.

He told me about his cousin's sighting in '51 or '52. Cousin had told only Dick because he was scared of community derision, he did not want to be the laughing stock! He had sworn Dick to secrecy but Dick's cousin is now dead, and Dick told me this simple yet astounding account.

Dick's cousin's hay field was five miles from where my brother and I lived at the time on our small family farm; the UFO landing was outside of Toeterville, just up the road. It apparently happened in the same season (autumn) as my first encounter, same year

Odd coincidence that Dick stopped by my place in 2011!

Anyway, throughout 1951 and 1952, my mother warned me to stay away from my brother's bedroom at all costs because he had armed himself with a baseball bat and large knife, against "three little men" he called Moe, Hoe, and Poe, or if you prefer, Mo, Ho, and Po.

My parents just laughed about this series of nightmares my brother was having with three little men, but his dangerous reaction was no laughing matter. He was always ready to attack. How could a 14 or 15 year old have been so confused between nightmares and reality – as to arm himself in the waking world and remain in attack mode?

In his adult life, my brother has no interest in UFOs or aliens. He is one of those people who figures there is life out there in the universe but it probably never visits Earth. He did have one spectacular UFO sighting while hiking with his (adult) son near Las Vegas. Tragically, his son was killed soon afterward.

My theory: My 14 year old brother, for whatever reason, was incompatible with the aliens' energies. I have no idea if he encountered them in the dream state (nightmare state) which the aliens might enter (?), or if these were real abductions which his mind, or the aliens' hypnotic suggestion, placed in "nightmare land" because he simply could not handle that it was real, happening in the waking world.

At any rate, Dick stopping at my house out of the blue in 2011 and telling me of the Toeterville saucer landing, is the near-proof of UFO activity at the time of my encounters.

Next, proof of UFO activity at the time of the Canadian cabin encounter, but knowledge of it manifesting many years later out of the blue:

On that Canadian vacation, my family met up with my Uncle Bill and his wife Marge and their two sons. They stayed in a separate cabin. As the years passed after the Ontario vacation, Bill and Marge got divorced and Marge did not keep contact with my family since Bill was our blood relative.

However, one Christmas, Marge sent a brief hello to my mother and they then exchanged one phone call as a "catch up" over all the years that had passed. When Marge asked "How is Diane?" my mother said, "Ah, she's interested in UFOs."

And Marge said, "You know when Bill and I were out in the boat on Eagle Lake that time, we saw two UFOs."

When my mother told me about this, chills ran up and down my spine, because I remembered that encounter on the Eagle Lake vacation but had never mentioned my suspicions to anyone or really pursued remembering. Our families only stayed the one full day at Eagle Lake and so this must have been on the same day as my encounter in the cabin!

When I found this out after Marge's phone call, I was around age 32. It had never been mentioned before, and there is no way this influenced my partial memories of the Canadian encounter. They were already there in my head.

Memories surface like rubber toys pushed to the bottom only to bob back up to the surface.

Odd coincidence that Aunt Marge phoned and talked about this after many years!!!

There is a pattern to my two encounters and their unfolding which might be helpful in other cases:

The pattern:

1. The encounter happens.
2. Evidence is left behind, apparently on purpose: The lost frenulum (#1 encounter) and the "larded" Little Golden

Book (#2 encounter). The book was about not fearing the unknown.

3. Someone comes out the woodwork many years later, and somehow feels motivated to mention UFO sightings/landing right in the area, right at the time, of my encounters.

One more thing, as Columbo says:

I was an only child of older parents. My mother had several miscarriages and was told after ovarian cyst surgery that she likely would not have any other children (she had had a child 10 years before, my half-brother Norm from an earlier marriage). Ten MONTHS after her surgery, I was born.

I am primarily left-handed but am cross-dexterous, using my right hand for some things, my left hand for others. Only 1% of left-handers are female. 10% of the population is left-handed. (cross-dexterous is an even less percentage). I have red hair, which is going extinct, so they say, in less than 100 years. So if one looks at the long odds of my particular genetics ever happening, plus my "miracle baby" arrival, I do have to wonder if UFO occupants sometimes "tag" people from birth or even conception or somehow cause recessive genes to manifest. (Wild speculation).

Diane Tessman is the author of such books as: *"Future Humans and the UFOs: Time For New Thinking"* and *"BEINGS FROM BEYOND: They Are Here"* available from Amazon.com

14.

MY MEETINGS WITH LIEUTENANT COLONEL JESSE MARCEL

By Calvin Parker

Editors Note: I know that earlier in this book I stated that we wouldn't be including much about alleged crashed UFOs, especially Roswell. However, this first-person account from Calvin Parker, which was originally published in 2020 for the Australian *UFO Encounter Magazine,* hasn't received a lot of publicity since it first ran. I think that Calvin's account is so compelling that it deserves as much attention as possible. We've seen this scenario before in UFO history, and it is a perfect example on how possible evidence can easily slip through our fingers unnoticed.

• • •

Most of you will know that my involvement in UFOlogy is the close encounter I experienced with Charles Hickson in 1973.

I also had a "missing time" episode in 1993, and a couple of bizarre events when I was just a kid. All of this has been covered in my two books *Pascagoula – The Closest Encounter*, and *Pascagoula – The Story Continues*. Both books were published by Philip Mantle at FLYING DISK PRESS.

When Charlie's and my encounter in 1973 hit the headlines, I did my best to keep out of the way. Charlie was the man in the limelight and that was fine by me. Charlie met with UFO researchers, other witnesses and lectured at UFO conventions etc. I was content to try and get on with my life, but occasionally I would pop up for one reason or another, or I would see or hear of another story in a newspaper about me.

As a result I did not meet that many UFO researchers or other close encounter witnesses, nor did I know or want to know much about other UFO events or sightings. However, in the early 1980s I met an elderly gentleman that is known to many because of his direct connection to the alleged UFO crash at Roswell, New Mexico in 1947. That man was no other than Lieutenant Colonel Jesse Marcel (USAF retired).

At the time, I was working in the oilfields and I had been transferred to Golden Meadow in Louisiana. I had been living there for about six months when I met a young lady by the name of Cathy Brown (real name on file) who worked at a local grocery store. I had not known her that long when out of the blue she asked me, "Are you the Calvin Parker that was abducted by aliens?"

I didn't really want to talk to her about this so I tried to change the story a little bit. I asked her how she knew about what had happened to me. She said that she had read an article about me just a short while ago and wanted to talk to me about what had happened. I had no idea about this newspaper article and to this day I've never seen it. It shouldn't have surprised me really as it was not uncommon for some dumb journalist to write a story about me and Charlie when they had nothing better to do.

Cathy informed me that she knew a gentleman that wanted to meet me and talk to me. She said he was getting on in years and he wanted to get something off his chest. She told me that this elderly gentleman was called Jesse Marcel.

I have to be honest and tell you that I had never heard this name before, and I had no idea who this old guy was. I was not interested in UFO stuff and the only other person I had met that you will know of is Betty Hill. I had been to see abductee Betty Hill and spent a few days with her. I believe Charlie Hickson met her at some point as well. I only went to meet Betty in the hope that she might be able to shed some light on what had happened to me and Charlie. She was a lovely lady but she was as much in the dark as we were.

So, I agreed to meet Mr. Marcel not knowing what he wanted, but if he just wanted to talk or if he needed some help then I would be more than happy to oblige. I know only too well what it's like to need help and not to have it.

MEETING NUMBER ONE

At this point I had to get back to work, but two days later I was back in the grocery store and saw Cathy again, and she asked me if I had considered the request to meet Mr. Marcel. She told me that he still wanted to talk to me and was very excited at the prospect or our meeting. Cathy told me that she got off work that day at 5:00 PM, so I told her that I would pick her up after work and we would drive out to Houma, Louisiana, to visit this old boy. It was only about a twenty minute drive from where we were.

My plan was to be polite, go and have a talk for a few minutes, and then make polite excuses and leave. I had no idea who Mr. Marcel was. I suspected he might just be a lonely old guy who wanted someone to talk to.

When we arrived I parked the car in the front yard, and Cathy got out of the car and walked in through the front door without knocking. This made me feel that she must either know Mr. Marcel very well or he was in some kind of trouble.

I nervously followed Cathy into the house and Mr. Marcel was in bed. He said he wasn't feeling very well and introduced himself.

We were not there long when Mr. Marcel proceeded to tell me the wildest story I had ever heard at that time (apart from mine and Charlie's and Betty Hill's). He told me straight up that a UFO had crashed and the US government had tried to cover it up (at Roswell, New Mexico in 1947). He claimed that the government gave out fake information of where the UFO crash site was so that no one would know where it actually happened. He then went on to tell me that some kind of special military troops were moved into the area to pick up all of the debris from the crashed UFO.

Mr. Marcel, to my amazement, was part of the military at Roswell and was the first military man on site.

At first he said he was allowed to talk about what had happened, but later was told not to say a word in fear that the Russians might find out. He told me that he was ordered to say that it was just a weather balloon that had crashed, and, being a good soldier, he carried out those orders. Mr. Marcel started to get a little tired and said that he'd like to talk to me again but felt uncomfortable talking in the house, so I arranged to meet him again in a couple of days at a nearby motel.

After leaving Mr. Marcel and heading back home I still didn't know much about who he was. I decided that it might be a good idea to phone someone who might know a bit more about old Mr. Marcel. The only person I could think of to call was Charlie Hickson. We hadn't talked for some time but he was the only person that I thought might be able to help.

So, I phoned Charlie and sure enough he knew of Jesse Marcel and wanted to know more. Charlie told me that he wanted to meet him the next time we talked, but I told him I would have to get permission first. So, I asked Cathy to call Mr. Marcel and ask if Charlie could join us the next time we met, but the answer was no. For some reason Mr. Marcel trusted me but not Charlie. I have no explanation for this; it just is the way that things went.

MEETING NUMBER TWO

The second time I met with Mr. Marcel was at a motel not far from his house. The Ramada Inn I believe it was. We met in a small conference room there. This is where he told me that the US Air Force didn't lie about the weather balloon story.

He said there was indeed a weather balloon involved just as they had said, however, he told me that it looked like the UFO (craft) had somehow become entangled with a weather balloon and crashed. He said there was debris everywhere.

Mr. Marcel told me that he didn't see any bodies and that a special team was brought in to clean up and recover all of the debris, and to do some kind of recon on the craft.

After a close inspection at the crash site, the military loaded everything on to trucks and took it to a military base and hid it in a hangar. They had guards at the hangar with orders not to let anyone in other than those members of the special team. He told me that they sent people out to speak with the local authorities to see how much they knew about what had happened and told everyone that it was just a weather balloon. It was at this point that this second meeting came to an end.

MEETING NUMBER THREE

I have to be honest and say that I now wanted to know more. So I dropped by Mr. Marcel's house unannounced and soon found out that this was the wrong thing to do. I could tell right away that this upset Mr. Marcel. However, I made my apologies and we sat and talked for a few minutes, and he started to talk again, but not about the UFO crash. Although he did tell me that the Air Force flew all of the debris from the incident out to another military base and he never heard any more about it. I asked him if there were any alien bodies, but he said he hadn't seen any bodies, alien or otherwise. But he was in no doubt that this craft was unlike anything he had ever seen before and he was ordered never to talk about it.

One intriguing thing he told me is that he did save three pieces of the debris from the crashed UFO, and that it was like nothing that he had ever seen in his life. He said that it wasn't anything of this world. I asked him if I could see these three items and at first he said yes. They were hidden in the top of his hot water heater in his house. All you had to do was to undo the top two screws on the water heater and remove the lid.

Just as he told me this there was a knock on his door and there was someone wanting to talk to him. Minding my manners I told Mr. Marcel I'd leave him to speak to his caller and that I'd see him and the items in the hot water heater at a later date.

Unfortunately I never got to see him again, nor the items that were hidden in his water heater. Working in the oil industry I was given instructions that night that I had to leave on a job that was off-shore, and I was gone for 154 days. By the time I got back I heard that Mr. Marcel had passed away. I never got to see Cathy again either but if I'm honest I didn't really look for her either. We were only friends at the end of the day.

All I can say is that in my humble opinion Mr. Marcel, or should I say Lieutenant Colonel Jesse Marcel (USAF retired), was an honorable man who had served and loved his country. The events that he was recounting to me are of course about the alleged UFO crash at Roswell, New Mexico, in 1947. I have tried to recall this as well as my memory will allow.

I have only decided to tell this story now as I just recently found out from my friend Philip Mantle that Mr. Marcel's son, Dr. Jesse Marcel Jr., had passed away a few years ago. I know how journalists can hound people so I kept this story to myself pretty much until now.

I cannot of course confirm any of what Mr. Marcel told me and I wish I'd had the chance to look into his hot water heater to see what was hidden there, but I didn't. I have recalled what we talked about to the best of my recollection and it is up to you whether you believe me or not.

One thing did occur to me, and that is if Mr. Marcel's house is still there, then might the old hot water heater still be there as well? Just a thought.

ABOUT THE AUTHOR:

Calvin Parker is a close encounter witness and the author of two books about the 1973 Pascagoula UFO abduction case.

He can be contacted via his website at: www.calvin-parker.com and at Flying Disk Press:

www.flyingdiskpress.com

Originally Published in: *UFO Encounter Magazine* - Issue #309 - Feb/March 2020

Calvin Parker

15.

TIM BECKLEY'S Q AND A WITH TED PHILLIPS

Ted R. Phillips was the Director of the Center for Physical Trace Research. He began investigating UFO reports in 1964 and was a research associate of Dr. J. Allen Hynek from 1968 until Dr. Hynek's death in 1986. It was at Allen Hynek's suggestion that he began specializing in physical traces associated with UFO sightings in 1968. Ted personally investigated some 600 UFO cases.

He was a member of a select team invited to meet with the United Nations Secretary-General at the UN in New York, along with Hynek, Jacques Vallee and Gordon Cooper. He gave two presentations at the First International UFO Congress and a presentation at the first MUFON symposium. Most recently he was a part of the History Channel documentaries "UFO Hunters" and "Alien Encounters."

Phillips died in 2020 at the age of 78.

Ted is here interviewed by the late Timothy Green Beckley, whose friendly, open and breezy demeanor are on display here, along with his sense of humor, a quality for which Tim will always be missed. The interview is followed by a transcript of Ted's lecture, which provides more detail on the alien artifact touched upon in their conversation below here.

Beckley: Well, here we are again with another edition of our Unfair, Unbalanced, and Unedited Report. Today I've got my old buddy here, Ted Phillips. And we're going to go on a journey you're not going to believe. So let me take my position behind the camera, because remember I can see you but you can't see me.

So, Ted, you and I did a wonderful interview when we first met several years ago. And we've stayed in communication. We say hello on Facebook as our photos pass by quickly. And you post some photographs of you and Dr. Hynek. You've been at this so long. You were so young. Just like me. I should talk. I was like fifteen when I started this. Were you about that same age too?

Phillips: Actually I got interested in this stuff when I was nine, but didn't do any real work until about twenty-two.

Beckley: Well, what was your first case?

Phillips: Socorro.

Beckley: Socorro! Yes, there's a photograph of you – is it with Lonnie Zamora and Dr. Hynek, and you're around the burn mark in the ground? Were you living in Missouri then and you just got up on your own and decided –

Phillips: Yes.

Beckley: Dr. Hynek didn't know you were coming? He didn't invite you?

Phillips: I didn't know Dr. Hynek at all, no.

Beckley: Now, talking about Socorro, somebody recently came up with this totally ridiculous theory that Socorro was like a hoax created by some college students with a balloon and a projector or something. In one or two words, what do you think of that?

Phillips: I can't really say it.

Beckley: Just remember, this is unfair, unbalanced and unedited.

Phillips: Bullshit.

Beckley: There you go.

Phillips: I have seen, over the years, folks come along with really, really dumb, uneducated ideas to explain away Socorro. It's a very complex case, with a lot of aspects to it and a lot of points that tell us it happened. And physical evidence and so on.

Beckley: Actually, the witness, Lonnie Zamora, just passed away within the last couple of years I think. He seemed pretty sincere to you?

Phillips: Yes.

Beckley: I guess he did to almost everybody.

Phillips: Yes, I was fortunate to talk to him when the media hadn't gotten to him yet. So in 1964, I got to talk to him right after the event. Then a couple of times through the years later.

Beckley: Did anything change about what he was saying? Or did he come up with some different beliefs or theories?

Phillips: No. It was kind of the "Jack Webb" thing, which fits a policeman. "Just the facts." And I was very impressed.

Beckley: I guess most people don't realize, but there were other witnesses to that. There was a couple – didn't they stop at a gas station and say they had seen something unusual? Were they credible? Did anyone interview them?

Phillips: No, I don't think they were interviewed. It was the word of the gas station attendant. But nevertheless, this was before this jumped to a higher level. And no one really cared. No one really knew anything about it. But there was also a family. I'm thinking they were from the Chicago area. Ray Stanford, of course, did the great book on Socorro. By the way, he asked me about a year and a half ago, "How about you and I doing a paper on Socorro together?" And I said, "Absolutely." And of course I haven't heard from him. But he's a busy guy. He's digging up dinosaurs.

Beckley: That's what I hear. So now, what other cases, just to briefly – because we're somewhat limited here today. We've all been hanging out at Pat Marcatillio's 51st – can you believe that? Fifty-one conferences. That almost boggles the mind. It does. Just rounding up the speakers and he's got everyone that he's recorded over the years. Just about everybody. Today Edgar Mitchell was on a live screen here. And of course you spoke. I think this is the first time in a long time that you've been out east here, right?

Phillips: Yes, actually it is. Last year I went through some health problems that were fairly significant. But I'm doing great, doing much, much better now. I have several lectures lined up.

Beckley: Wonderful. Because I know people want to hear what you have to say. Especially with some of these new things that you've just started talking about. Marley Woods. And even more exciting to me, because of my interest in the legends and lore of the underworld, is your future trip coming up pretty soon we hope to Slovakia. And, in three minutes or so, just give us a little hint about what that's all about. I know you've been there several times before. There's something that's ticking or making a sound?

Phillips: (laughs) Yeah, a ticking bomb. Actually, a Czech engineer, who had three engineering degrees, during World War II, was being hidden in a cave after his battalion was wiped out by the Germans. And they were hidden there by a Slavic sheepherder. This was a cave that he held in very high regard. He kept the entrance covered with stones and had been taken into it in the 30s by his father and grandfather.

Beckley: What year was it discovered?

Phillips: it was discovered in October of 1944. And the Czech engineer had been captured by the Germans, put in a concentration camp in 1939, escaped in 1941, which didn't happen too often. He joined the Slavic underground army and became a captain. He'd fought in World War I so he had a lot of combat experience. So they were under attack and they were all killed except for this gentleman and two of his soldiers. And the sheepherder hides them in this cave, which is a cranny that opens into a very large, expansive room.

And immediately he notices at the back a passageway that went on. They had not had food for four days and bat sounded really, really good.

Beckley: I eat it all the time. (laughter)

Phillips: But the Slavic gentleman asked him not to explore the cave, not to go back, because it was dangerous, there was gas and bottomless pits. And it was haunted. One of the interesting things, when he first took the Slavic soldiers in, is that he went through the religious rites, blessing the guys, himself and the cave. It was a very significant thing in his life. And he had been taken to the back of the cave and obviously had seen the artifact. And I found out much later, in 1971, that he actually had admitted to knowing this thing was in there.

Beckley: Now, when you say "artifact," it was making a sound?

Phillips: Yeah. A pulse that would affect the tri-meter, three compasses that I put in the center of roughly a 25-foot area that I ran on to trying to find the entrance. And batteries on all the equipment go dead when you get in there.

Beckley: Well, maybe it's working on my camera today.

Phillips: Yeah. It's not that far. Seven thousand miles. But anyway, being quick of mind, I thought, well, if the tri-needles are being activated by something – so I put the tri-field and the three compasses out in the middle of this area. The camcorder on a tripod. I backed it outside of the affected area so it would run, and zoomed in with the telephoto on the instruments. And great video of all these instruments, the compasses and the tri-field. Every two seconds, pulsing, dropping off, pulsing. The tri-field and the compasses pulsing back and forth every two seconds.

Beckley: Which would indicate what?

Phillips: Well, it was indicating some kind of, I would think, an EM sort of –

Beckley: Is someone or something keeping something alive?

Phillips: I don't know. But when the engineer –

Beckley: It almost sounds like a real-life version of "The Thing."

Phillips: It could well be. But remember I'm going into this place, so let's not get too graphic. But this artifact, because of its surroundings, and the situation it was found in, is extremely old.

Beckley: How old do you think it is?

Phillips: Well, based on the limestone drip-age on an average, and the effects of flowstone, stalagmites, growing on the face of the outer wall, it would have to be far in excess of 6,000 years.

Beckley: Well, how in the heck did it get down there?

Phillips: That's a good question. From the cave entrance –

Beckley: It's not big enough, though, the cave entrance, right?

Phillips: Oh, no. And they found the prehistoric cave bear inside this thing. And the cave bear could not have gotten through this small crack leading to the interior.

Beckley: So maybe the Earth has shifted?

Ted Phillips

Phillips: Could be. It definitely has been there for a very long time.

Beckley: Or there could be another entrance?

Phillips: He couldn't find any. Now he didn't have the time to explore the many side passages. To get into the bigger room, where the artifact is at, the cave bear would have had to crawl through a two-foot high space and a cave bear couldn't do that. He could barely get a foot through there. They were massive, massive dudes.

Beckley: Hopefully you won't find one of them hibernating.

Phillips: Not likely.

Beckley: Okay, so you have a documentary deal or a movie deal on this? We're finding out about it soon?

Phillips: Yeah, next month.

Beckley: Steven Spielberg?

Phillips: No. Probably some escaped convict. That's who I generally work with. (laughter)

Beckley: Whoever's got the money.

Phillips: Yes, absolutely.

Beckley: So what are your chances of finding this thing?

Phillips: I think pretty good. As a matter of fact, I think I have the right cave and there are a number of indicators I mentioned in the presentation that are pretty valid stuff. And not only that, the lay of the cave, I have the original cave map as drawn by the engineer in 1944. And he kept all this, all the sketches.

Beckley: I've seen the documentation. Today you gave an almost two-hour presentation and that was just fabulous.

Phillips: People were actually even awake.

Beckley: That certainly puts you out of the realm of Socorro.

Phillips: Yes.

Beckley: And you had told Dr. Hynek about this and he said go ahead, go for it?

Phillips: Well, actually he came out right away to the engineer's home in 1970, and we both visited with him and his wife.

Beckley: They were living in Denver.

Phillips: in Boulder, Boulder, Colorado. And very, very sincere people. And they had been through a lot. First the Nazis and then rescued by the Russians and turned on by them. So they were living the life of leisure prior to the German invasion and that went to hell in a handbasket. And being thrown in a concentration camp, that's –

Beckley: So what do you think this is? Is it an ancient alien artifact?

Phillips: I really have no idea. I really don't. There's no indicator other than it obviously was constructed of some kind of material we're not aware of today. A seamless, large device like this and a jet-black mirror finish material. And he actually used his military rifle and shot at the wall because he couldn't get a piece of it for a sample. And the bullet didn't even scratch –

Beckley: Now, is this a craft or a container? Is Superman inside?

Phillips: He may be. Only Superman could move this thing. It's very large.

Beckley: Well, I have followed you for several years talking about this and I find this probably one of the most intriguing, ongoing incidents.

Phillips: Like the old Shaver stories, it never ends.

Beckley: One might find it hard to believe that an associate of Dr. Hynek would be investigating other realms and other theories outside of spaceships from Mars. But I know for a fact – I spoke with Dr. Hynek just before his passing and he told me some very interesting Men-In-Black stories. I found that intriguing. I had taken a photograph of what might have been an MIB, I don't know. But you have moved, as have several other people in the field, from a very conservative approach to the subject of UFOs or flying saucers to someone who is now investigating or thinking about parallel realms and other dimensions. What brought you along this path? Have you given up the idea for good that these may be spaceships?

Phillips: I have. And actually J. Allen was the guy that converted me. I didn't know it. But he quite often said the extraterrestrial theory is far too simple. And he said we have to go far beyond that. He said this is much more complicated. And I was grubbing around for nuts and bolts or oil leaks.

Beckley: Well, you had over four or five hundred cases of landing traces?

Phillips: Back then, yeah. Now I have 5,189.

Beckley: Are they still taking place?

Phillips: They are, but at a very reduced rate. And by different types of devices. One thing that I noticed, in the late 80s, we started having less landing cases involving, say, a thirty-five foot metal disc, with little guys running around under it.

Beckley: I miss those little guys.

Phillips: But what has replaced those are the much smaller light balls, ranging from baseball to beach ball size.

Beckley: Orbs?

Phillips: Well, I don't like to call them "orbs," because of light and dust in photos of "orbs." These are actually things that you see, visually, and they have a tendency to come quite close in some events.

Beckley: Well, I know I just purchased from you this afternoon a copy of a DVD you did on Marley Woods, which is an incident there in Missouri, your home state, involving these mysterious – I guess I would call them "ghost lights," "spook lights." What do you call them?

Phillips: Actually, it runs the entire spectrum of UFO-type events. Now, even into what are typically called "paranormal" sorts of events. So it comes down to this – and I have the cases to back it up – if you have a blue light ball the size of a baseball and it's flying over a field, you can clearly see it, you can capture it on HD video, you have what is typically a UFO. What if you see precisely the same thing inside a house? Is it then still a UFO? Or is it a paranormal event? Because a lot of paranormal researchers see and capture just that kind of activity. And there are good cases ...

Beckley: Are you familiar with the case in Colorado's Black Forest? Do you know that one? I believe the guy's name is Steve Lee, who has all these objects coming out of the wall and out of the mirror and outside the house. There are a few things posted on YouTube, but not many. It was on "Strange Universe." All kinds of things just shooting around the house and coming out of the woods and stuff like that. It was another one of these cases and it seems to be really, really perplexing. He even wrote to his state senator who came up and saw some phenomena and they heard voices and it's a whole mixed bag of paranormal activity.

Phillips: I'm not familiar with that. It sounds sort of like what we're having in the Marley area. So, at any rate, the way I look at it, no matter what your prior thoughts might be, if you start running into something new and different and it continues, and if a video camera can catch it, you can be pretty sure it's something kind of physical anyway. And you have to go in that direction. I'm not giving up on the nuts-and-bolts or the urinating thing, but ...

Beckley: UFOs urinate?

Phillips: No, the little guys. You have to look very closely.

Beckley: Have you really found cases like that?

Phillips: No.

Beckley: Oh, too bad. So do you have any conclusions at all about what's going on here?

Phillips: I really don't. I hate to say that, after 48 years. But I'm even more convinced by the data and the evidence that something very real is going on and it's just beyond our knowledge at hand. The only way to come to a conclusion is to get enough data to come to a conclusion.

Beckley: Can you come to a conclusion?

Phillips: No.

Beckley: I don't mean now, but even in the future, is it possible to come to a conclusion?

Phillips: I'm not sure.

Beckley: How do you prove the conclusion?

Phillips: That's a very good point, and I think about that a lot. Why do I do this? Why do I continue to do this? Except it's fascinating.

Beckley: It's fascinating. This is what I tell people, that I don't know if it matters whether somebody has told a tall tale, if you want to believe it. That they went to Venus or Mars. It's all part of the same package. I kind of feel like we're all on the mother-ship being drawn somewhere for some reason, or maybe no reason. But we're all kind of in this together, whether we like it or not, so I usually keep my sharp tongue in my mouth. Unless I happen to be doing a vampire movie or something like that. Then I can perhaps let my fangs show.

But I just figure if somebody's onto a trip of their own – now, mine happens to be synchronicity. Patterns and coincidences that happen that are just not even explainable. You can't have even a remote reason to say that this would happen. And people know about this. I've talked about it on the air and written a couple of articles and so forth. Are there any other cases that you're investigating currently?

Phillips: The Marley stuff is working towards a cave trip. That's primarily what I'm doing. The Marley things keep me plenty busy. It's remarkable. I have the opinion that if one could observe the entire area there that you would pick up something every evening. And it's seen in daylight also.

Beckley: Is it? What is the appearance that they take on?

Phillips: Pretty much the same as at night. And the light balls have been seen in daylight, the big amber's been seen and photographed in daylight. The problem with this stuff, Tim, is it makes no sound. But one evening one of the team members says "Behind you." And I turn around and about six feet behind me is a baseball-sized light ball. Not very high above my head.

Beckley: Well, it's your halo.

Phillips: Yeah, sort of. But you'd have no idea it was even there.

Beckley: There was no warmth? No tingling sensation or anything like that?

Phillips: No.

Beckley: Do you think you could put your hand through it?

Phillips: I would not try. There's a good deal of –

Beckley: It's not ball lightning, though?

Phillips: Oh, no. It's not attached to anything. Observations have a duration of a half hour to an hour. It's not something...

Beckley: Now, some of them have taken on the appearance of craft?

Phillips: There have been just a handful of solid structured objects.

Beckley: Well, one is enough. There seems to be something to that. Otherwise we could say that they're ghost lights or orbs or something and might not be UFOs. But if they do come in a structured craft – unless maybe THEY'RE coming to look at the lights.

Phillips: Yeah, there's the occasional UFO.

Beckley: What's the largest light thing that's been seen there?

Phillips: Well, the largest object would have been two structured objects that were captured on video, estimated by local witnesses to be about the length of a football field.

Beckley: While I'm thinking of it, why don't you tell people where they can get your DVD?

Phillips: That's really a good question because I shut my website down because I got tired of people emailing me and saying "You need more photos." And I thought, why am I bothering with this website? So I guess they'll just have to catch the lecture. So much for a website.

Beckley: You say there's no heat, there's no sound. How many witnesses have there been now to the Marley Woods phenomena?

Phillips: In Marley, 227. And not a single person has gone public. And you don't make up a big, fat lie and then keep it to yourself.

Beckley: Are these all people that live in that area?

Phillips: Yes.

Beckley: No casual observers have come by or you haven't invited others?

Phillips: Well, I do have a few of those types of sightings but generally I'm not able to get names because they don't want any part of it. The people "in the woods," so to speak, trust me, and I've protected them for fourteen years. So they have no doubt that I'll continue to.

Beckley: You don't think that there is any possibility of a hoax?

Phillips: Oh, Lord, no.

Beckley: Did I hear you describe, on some show, something to do with a dog or a mutilation?

Phillips: Well, there have been some rather severe cattle problems, which before this I didn't pay a lot of attention to because I thought it was coyotes, hungry animals and things. But I'm finding that probably is not totally true.

Beckley: But wasn't there something to do with dogs?

Phillips: Well, this borders on the paranormal. It's not like a little guy gets out and steals the dog and eats it. It's some rather extraordinary events that are under investigation –

Beckley: But what was the story on the dogs? They got caught on a fence or something? I'm trying to remember –

Phillips: It would take a long time to get into that. It is a very protracted – two different situations – really it involves a lot of stuff and it's a devil of a story, really. If we had an hour, we might get into it.

Beckley: Well, can people order the DVD from Amazon?

Phillips: No, not yet. I just put this first DVD together, which is made up of videos of the various activities. And part two will be more of a story with the investigations and a lot more content. And maybe a little less video. But the two combined I think would be of great interest, really. There is a lot of interest in this particular event. And one thing that I would throw in — it is, in a lot of ways, very similar to Skinwalker Ranch. And there are a couple of other areas like this.

Beckley: Well, there's the Bradshaw place that I recently kind of got involved with. I was down in Sedona and interviewed this fellow Tom Dongo, who's kind of the main researcher of this particular place. There's all sorts of mysterious lights and things showing up and interdimensional figures being captured – whatever they are – on the photographs. And finally the government ended up buying the land. So now it's government-owned. People do camp around there but it seems to go in waves and there hasn't been anything in recent –

Phillips: Well, the government is probably going to open a park, maybe a waterslide park.

Beckley: It's been vacant now. I met the woman, she sold it to the government. They offered her a nice price.

Phillips: The government moves slowly. You don't want to sit out there by the edge of this land and wait.

Beckley: I think it's had a number of trespassers. The fence has been cut through and all that. If they're reporting anything, they're not coming back to talk about it.

Phillips: There's that too. At my age that's kind of exciting, the possibility of not coming back.

Beckley: Wow. Well, you've got a lot of work to do.

Phillips: I mean, I don't want to drop dead in my own house.

Beckley: Nor in a cave, I wouldn't think.

Phillips: Well, now, caves are okay. I love caves.

Beckley: Have you done a lot of spelunking?

Phillips: Oh, my Lord. Since I was fourteen.

Beckley: So you're seasoned at doing this. It's not like you're some novice who's going to –

Phillips: I belong to the National Speleological Society.

Beckley: Have you ever written anything or corresponded with any people who do this? There must be serious cave explorers as opposed to –

Phillips: I'm a serious cave explorer. The longest cave I've explored was 20 miles and, believe me, that's pretty serious.

Beckley: I would say so. I'm just thinking that whenever you ask someone who does spelunking, questions about voices and things that they might have seen or heard, a lot of them look at you like you must be –

Phillips: They look at you that way no matter what's going on.

Beckley: But there have been numerous cases, and I've written about them in some of my books, of weird things, doors opening down there and people seeing beings. Some of them that have even been led to safety by whatever. And of course we have the Tommy-Knockers.

Phillips: Yes.

Beckley: In fact, you mentioned Colorado. There's a restaurant, I think it's outside of Boulder, or maybe Denver, I can't remember which. It's called the Tommy-Knocker and they've got little carts full of fake gold and Tommy-Knocker paintings all along the wall and Tommy-Knocker beer.

Anyway, Ted, this has been an enjoyable conversation and quite a weekend. And you made it happen here at Pat Marcatillio's 51st UFO Conference in the great, grand city of Bordentown, New Jersey. That's rock and roll.

Phillips: Yes, thank you, Tim.

16.

TED PHILLIPS LECTURES ON THE ARTIFACT FOUND IN EASTERN EUROPE

The following is a lecture Ted Phillips delivered at the 51st annual UFO Conference hosted by Pat Marcatillio. In the lecture, Phillips provides a more detailed account of the alien artifact he briefly discusses in the previous Q and A with Timothy Green Beckley. Phillips makes references to some illustrations taken from his lecture which are not reproduced here, but hopefully the reader can still piece it together visually well enough to understand it.

• • •

In all the years I've been working this stuff, this is the single most important thing I've ever been involved in. This is Antonin Hurrah, a Czech engineer with four degrees in engineering, a couple of degrees in business management, spoke seven languages fluently. He was a very bright man.

In 1970 I received a call from an investigator with APRO, a friend of mine, and he said "You've got to come out to Denver and talk to my neighbor and listen to his story." Because this investigator knew that I was a cave explorer and into the physical evidence thing. So my wife and I went out and we talked with him extensively. He maintained, during World War II, a daily diary, and it's the best reading. And it's better than anything you ever read in very great detail.

This was his home, in the north Czech Republic, which was then called Bohemia. And this had been in the family for a long, long time. They had a lot of property. They had timber, they had uranium mines. Tony sold the first uranium specimens to Madame Curie in the early 20s. They were very wealthy people until the Nazis decided to take over Czechoslovakia.

In 1970 I was looking at these sketches in his home. And I photographed them directly from the diary. This is two pages. And I was so caught up in his story that I called Allen Hynek and he met Ginger and I in Las Cruces, and I showed him what I had. And then he flew up and talked to Tony. And we were so impressed that we decided we have GOT to mount some kind of expedition. And so through Jim and Coral Lorenzon, Jackie Gleeson agreed to fund my first trip over there. That would have been in 1970 and unfortunately the Russians had just invaded Czechoslovakia in 1968, and we had four contacts who would give me supplies and so on, when the Russians arrested all four and killed them. So that kind of killed the expedition. It was far too dangerous.

Now let me tell you very briefly what happened to this guy. When the Nazis took over, they took everything they owned, down to his wife's wedding rings, put him in a concentration camp. He was there from '39 to '41. He escaped in '41, escaped to Slovakia, and joined the Slovak underground army. He was then an army captain and he had a battalion of 184 men. They were in a heavy firefight with the Germans. All of them were killed except Tony, Martin and Yuri. And they were left for dead. Martin had very serious wounds.

A sheep man comes along the next day. He finds them and treats their wounds and builds a stretcher. And took them up into the mountains to his cave. The cave had a very small opening and it was closed up with rocks. The sheep man removes the rocks and they go in. It opens into a very large entrance room, and here he's going to hide them from the Germans. He starts going through the holy rites, blessing himself, the cave and everyone in it.

Which Tony found kind of unusual. There was an opening going on back at the end of this room, and he said, "Please don't go back in my cave because it is haunted and very dangerous." And so the Slovak left and Tony got his rifle, torches and carbide light, which I have out at my table, and went on back in the cave, which is what I would have done. After going about two miles in the cave, I'll show you through close-ups of his sketches, he's coming through here. And at that area he comes into a long, level corridor. At the end of the corridor is a crawlspace. He comes through there and this view right here, of this wall, this is 25 feet across the room. It's about 30 feet high. And this is the height of the man, to give you some reference.

So from this point, looking here, this is what he saw. And what he's seeing is exposed black, satiny mirror-like material. No seams, no rivets, no anything. And it is framed by very large cave formations. And if you know anything about caving, those formations take a few years to form. Right through here you can see a crack coming down, and where it meets the cave floor is just big enough – if he gets naked – he can go through the crack. And if you've ever been in a cave with all those rocks, you don't want to do that. But as he slides through the crack, he can measure the thickness of that outer wall at seven feet thick. He gets through the crack and he rolls down the floor to this point here into the back wall.

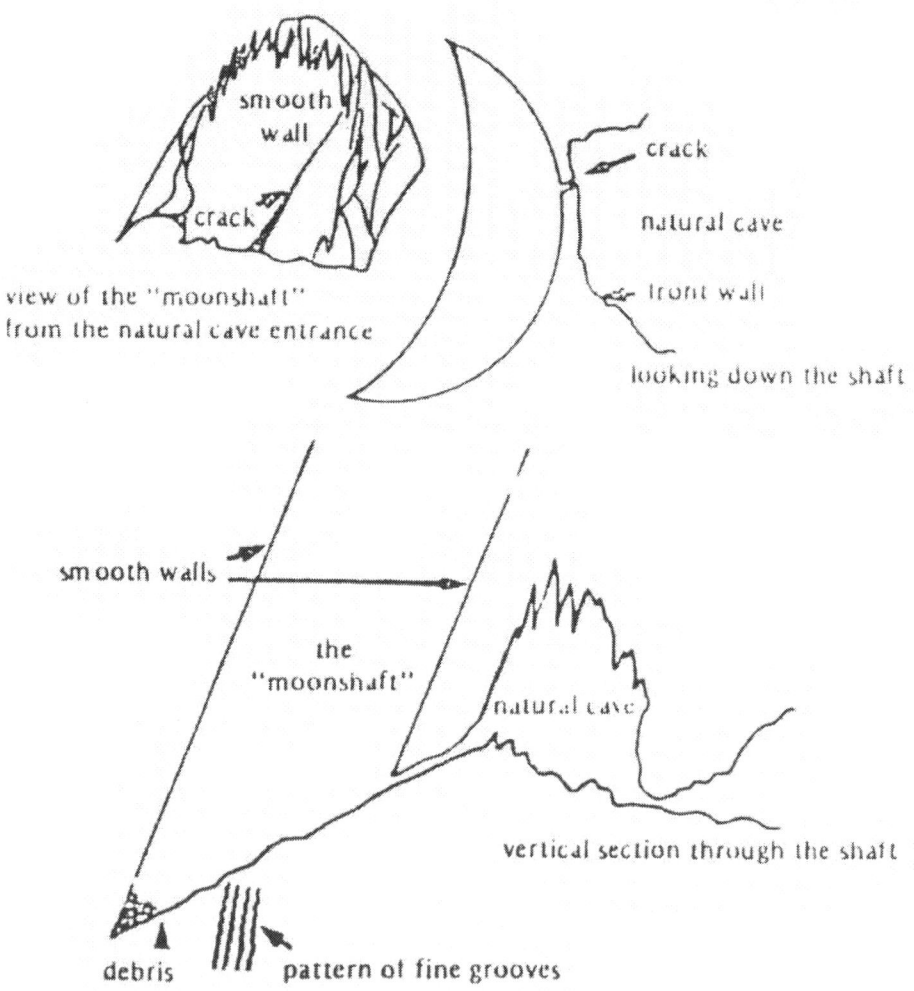

Figure 3. Schematic representation of the mysterious moonshaft.

And what he finds himself in is a large structure that rose up right here, the outer wall, the inner wall. So it's like a vertical shaft sixteen degrees off the vertical. He could never get enough light to find the top. He goes back to the camp to get more supplies, doesn't tell Yuri or Martin about it. And every day, he returns a couple of times to this – to survey it, to study it, to try to get samples. He fired his military rifle at this point on the wall to try to chip a piece off, and it didn't even scratch this material. He found, by digging right here, that this is a limestone accumulation, drippage. And he went down through five feet of this stuff, which means about 6,000 years. Beneath 6,000 years of limestone, he finds the skeleton of a prehistoric cave bear.

He takes three of the teeth, which I have. And after the war he went to a museum in the Ukraine, and the curator identified them as coming from a prehistoric cave bear. So what we're talking about is 6,000 years of limestone and a cave bear that's been dead at least 50,000 years. Perhaps a million. Beneath the cave bear, it's lying on a curved, wavy grill. He thought he could he feel heat so he put his ear and cheek to this grill and there was considerable heat coming up through it. And far, far below he can hear what sounds like a turbine engine.

And you've got to remember, this is an engineer. This guy was very familiar with subterranean sounds. So it becomes really interesting. What are we dealing with? This is how the thing would appear from above. That's the configuration of it. This is that outer wall, the crack and the room. And he said without question, these are constructed, mathematically curved walls. So he's sitting inside this thing, writing the diary and doing these sketches and pondering what it might be. And the last thing he wants to see is tomb-robbers getting in before scientists.

So on the cave map, which he gave me, when he came out the last time – they were going to rejoin their unit – and they had been in there for seven days – he decided that he would obstruct the cave passage where it became a crawlspace in three places so that tomb-robbers could not find the artifact if they did find the cave. And without the map you would never find this thing. As I say, fortunately I had the map.

This is Dr. Hynek and I in our meeting on the artifact. He then went up and had a long visit with Tony and they were both from Bohemia so they hit it off very well. My first trip over there – there's not a great selection of rental cars – this is going through village records. This is a manuscript about a crash of a circular object into a mountain very close to the mountain that the artifact is at in 1663. This is the area. It crashed in this area here. And the villagers from a tiny village went up and carried pieces of this thing down and buried them in the village. And I didn't know this on any of my trips over there so that's now on my to-do list. And guess who that is, struggling to breathe.

This is the area of the cave. That's the mountains a la Google. That's the cave area in the mountains. If you're standing at the cave site, this is what you see looking east, north, south and west. For a young, 45-year-old guy, it's a pretty good trip. Now this taken from inside what I believe may be the cave. If it's not this one, there are two more, and it's one of the three. It's taken inside looking out.

High moisture, which gives you all that fogging. Some collapse in the floor from aerial bombing in early '45. These are some carvings. Martin, the severely injured soldier, was placed in an alcove. Rocks were heated and placed under him to prevent chilling. And we found about 40 feet of 1940s bandage in there, which I brought back. We found this, and this is basically a calendar, and when you sketch that out it makes more sense. They went into the cave on October 23rd.

There are two more of these things, possibly. One in Oklahoma and one in Ohio. And they were discovered in 1867 and 1928. And this is the signed testimony of one of the witnesses from 1928.

I've been in touch, emailing, with two Russian scientists, and they found ancient manuscripts from Siberia where a large black cylindrical object with curved walls came out of the sky and landed. And this was eons and eons ago. It made a terrible noise. And it was so tall the villagers could see it from miles away. Each day it got shorter. So when it finally disappeared, the villagers hopped on reindeers or ATVs, whatever they had back then, and went to the area. There was no object, but there was a crescent moon-shaped chasm that went down and they have not found the bottom of it. Out of this is coming an electronic signal, pulsing every two seconds. They found another one in Yugoslavia, the same pulsing.

My second trip, when I was close to the cave, the compasses started going crazy. I put them on the ground with an EM field meter and videotaped them as they pulsed every two seconds. So there may be three of these things. And, believe me, it is a fascinating thing. So my goal is to get back over there before the blizzards start up this year and to help fund this, I am hawking CDs that I've made of all the unusual lights and cattle mutilations and so on. That's called "The Marley Woods." I have the ten best physical trace cases. I have the cave CD, a ghost light CD, and the Cato Landing CD. A full report, all the photos and sketches.

I've never been one to hawk or sell anything relating to this subject, because I think it's a little demeaning to the subject, but unfortunately I want to get back to this thing. So I thank you very much. Thank you.

(Applause)

Is the cave containing the moonshaft hidden within these mountains in the Czech Republic?

17.

THE MOST IMPORTANT ALIEN ARTIFACTS EVER DISCOVERED?

By Sean Casteel

The late podiatrist Dr. Roger Leir pioneered the study of alien artifacts over the course of the years beginning in 1995 up to his death in 2014 at 78 years of age. His 1998 book, "*The Aliens and the Scalpel*," details his early efforts in surgically removing what he believed to be alien implants left behind in the bodies of experiencers by the abducting aliens during the course of an abduction. The book is subtitled "*Scientific Proof of Extraterrestrial Implants in Humans*," and Leir himself felt he had discovered the "smoking gun" of ufology – hard, physical, scientific evidence of a continuing alien presence on Earth.

WHITLEY STRIEBER'S PRAISE FOR LEIR

In his Foreword to the book, Whitley Strieber, the author of the bestselling abduction account "*Communion*," sings high praises for Leir, writing that, "Sometimes the whole world knows when a historical figure makes his history. As often, though, people whose work is little known make history in the quiet of ordinary life. Dr. Roger Leir is such a historical figure. There will come a time when the extraordinary breakthrough that he has made is noted in every textbook, but right now most of the world has no idea of what he did or even that such a thing could be possible."

Strieber provides a short history of Leir's introduction to the subject of alien abduction. In June of 1995, Leir attended a lecture by Houston UFO researcher Derrel Sims at which Sims discussed the fact that many people who believe they have been abducted by aliens also feel that they have had strange objects implanted into their bodies. Afterward, Leir made Sims an offer: if witnesses would pay for their own airfare to Los Angeles, he would remove these objects free of charge.

On August 19, 1995, Leir removed objects from the feet and hands of three such people. The results are among the most remarkable outcomes recorded in the annals of surgery and are thoroughly covered in Leir's book. On May 18, 1996, another set of surgeries took place. This time, a large group of witnesses was invited to observe, which included Strieber himself. "To see these surgeries actually being performed was among the most moving experiences of my life," he writes.

Again, the outcome was the same: unexplained objects were extracted from the bodies of all three witnesses, which Strieber writes will one day lead mankind straight to the strangest and most provocative discovery of all time. Could somebody really be implanting objects into our bodies – somebody from another world? It seems fantastic to contemplate the idea that such beings could even find us in this vast universe. But it is even more incredible to imagine that they might not only be observing us, but also actually engaging in intimate intrusions into our bodies of which we have only dim memories.

IRREFUTABLE EVIDENCE

And yet the evidence is here, so overwhelmingly powerful that a sane person cannot easily deny that it is real, and yet so disturbing and challenging that even the best of scientists have trouble facing the fearsome questions that its existence poses.

Strieber believes that Leir should be acclaimed as a hero, having offered mainstream science the sort of data that it was certain that the world of UFOs could never provide. But Leir is still not a celebrated pioneer in the world of science because of what his findings imply. To accept it, we must also accept that someone is capable of entering our homes and introducing these objects into our bodies, without permission and in a way that is almost completely undetectable. Worse, we must face the fact that, although the materials that the objects are made of can be analyzed, we presently have no way of knowing what it is that the objects do.

All we have, according to Strieber, are the fragmented and confused accounts of people who bear the objects. Nevertheless, those reports are also fairly consistent. In general, witnesses who report objects in their bodies do not describe contact with wise or angelic beings. Instead, their visitors tend to be tough and determined, sometimes even brutal – pretty much what you would expect from aliens who are willing to carry out what amount to unprovoked assaults on their victims.

Strieber has himself twice experienced an implant being placed in his body. The first event, in 1988, led him to have a magnetic resonance imaging (MRI) scan of his head that showed a small, brightly returning object in the middle of his left temporal lobe. No matter what it was, there was obviously no way of exploring it or removing it surgically.

LEFT ALONE WITH THE MYSTERY

"So I was left in anguish about it," Strieber writes. "Did it mean my mind was being controlled by somebody, or perhaps monitored? Since it couldn't be removed, I would have to bear it. Did that mean that it would eventually destroy me? Might I get a brain tumor or suffer some sort of stroke? These are not questions I could answer, but that I had to bear without answering. I am not alone. Every patient who has presented himself to Dr. Leir has been in the same quandary. In fact, every abductee must deal with this issue, especially those who have possible implants that are visible from the surface of their bodies."

There has been no help at all for Strieber and other abductees from official agencies like the Centers for Disease Control, the National Institutes for Mental Health, from Congress or the White House under three presidents, or from friends in the military and intelligence communities.

"Nobody wants to face what appears to be an almost indisputable reality: people who report being implanted by apparent aliens actually have objects in their bodies that can be detected and removed," Strieber writes. "What more proof is needed than this? How could a human problem of such horrifying dimensions simply be ignored? It is almost impossible to believe, but any reader of Leir's book, upon seeing the way that science and the media reacted to Roger Leir's phenomenally provocative discoveries, cannot help but be shocked, even terrified. To me, as an abductee who is facing this up close and personal, it seems impossible to believe that the situation that now pertains could actually be true."

In addition to the implants removed by Dr. Leir, there are also now hundreds of hours of video of UFOs, some of it taken by the very sorts of professionals that the scientific community claimed would never see such things.

"So what are we waiting for?" Strieber asks. "Somebody is right here right now, and they are plunging deep into our bodies and our lives, and we are refusing to even entertain the evidence, let alone mount an effort to find out what in the world is going on."

LEIR'S EFFORTS AND THE OPPORTUNITIES THEY OFFER

Were it not for pioneers like Leir and Sims, this situation would be allowed to continue absolutely ignored, according to Strieber. As it is, Leir's untiring scientific curiosity has given us a chance that that we did not have before: instead of leaving this problem to future generations, he offers us a chance to look at the data and the struggles that surround it today.

"No matter how resolute science is in its denial of the obvious, or how helpless government and the military prove to be," Strieber writes, "at least close-encounter witnesses can now point to hard evidence that their experiences are not psychologically disturbed fantasies and dreams, but are instead real experiences that absolutely and urgently cry out for careful explanation and effective action. It proves that there is physical evidence of our encounters, and shows with authority that this evidence cannot be explained in any normal way. It sounds a clear call to action now, because what it reveals would appear to be the visible edge of a broad scale exploitation of the human species by unknown forces for unknown reasons."

But Leir has said "no" to the idea of ignoring the evidence for the close encounter experience and proved what was supposed to be impossible to prove. He went into a very dark place and turned on a light. "The Aliens and the Scalpel" is the story of how he did it.

Having agreed to perform the implant removal surgeries for free, Leir next sought help from a fellow medical professional he called only "Dr. A," in order to preserve his colleague's anonymity. Leir explained that he had lately gotten interested in Ufology, something Leir had not told even his closest friends at that time, mentioning also his acquaintance with Derrel Sims and how it all led up to the impending surgeries.

Dr. A. took Leir quite seriously, asking how Leir knew he could trust Sims and how he could count on the patients even showing up for the surgeries. Leir replied that he didn't know Sims intimately but was merely trusting his gut about the matter, as well as banking on his friend Alice Leavy, who ran the local chapter of the Mutual UFO Network and had great confidence in Sims and the legitimacy of his work. The two doctors agreed on the date August 19, 1995, and Dr. A. asked to review the medical data as well as preview the x-rays.

WHAT TO DO WITH THE OBJECTS?

Leir began to do intense research into what was then currently known about alien implants.

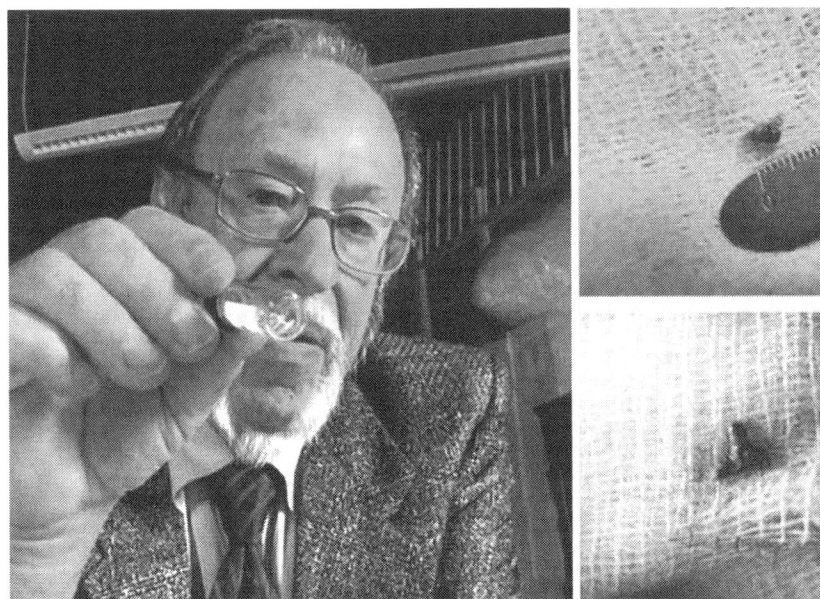

Dr. Roger Leir removed objects from people who claimed to have been abducted by extraterrestrials. These "implants" were found to have very unusual characteristics.

"Faced with tales of frustration and failure by others to obtain hard evidence of the reality of alien implants," Leir writes, "I was certain I had to take every precaution in order to prevent similar occurrences from happening to me. The scant literature indicated that a few objects that had been successfully removed had strange things happen to them, resulting in a complete lack of analysis. There were tales of surgically removed objects turning to powder, becoming liquid, vaporizing and other strange phenomena, which were equally as peculiar as the subject itself."

It was not enough to merely extract the objects, he explained. It was also necessary to preserve them in a way that would prevent them from rapidly degrading. After long and hard thought, Leir said that a simple answer dawned on him intuitively. He writes that this small insight, in retrospect, may someday be seen to be a significant turning point in the seemingly endless human quest for proof that we are not alone in the universe.

"The safest medium to place these specimens in," he realized, "would have to be a biological fluid belonging to the patient from whom the object as extracted. All I had to do was to withdraw blood from each one of the patients, spin it down in the centrifuge and separate the serum. The serum would be mixed with a preservative, and, without the presence of cells, it would not coagulate. This, then, would become the ideal transport medium."

CONSCIOUS RECOLLECTIONS PREFERRED

One of the criteria Leir and his team established for the removal surgeries was a detailed abduction history of each of the proposed surgical candidates, which Sims took responsibility for supplying. Due to the fact that critics and debunkers often point to the unreliability of regressive hypnosis as a means of obtaining an abductee's memories of an alien encounter, it was decided that the team would forego the use of regressive hypnosis altogether. Leir's books focuses only on two abduction histories from among the many surgical candidates he would work with, saying they are both typical cases and consist of as many details as could be recovered through conscious memories.

PATRICIA'S STORY

One case involved a young mother named Patricia, who resided in rural Texas with her husband and two sons. It was 1969 and Patricia was eight months pregnant with their third child. The couple decided to take a fishing trip to a place recommended by a couple of the husband's buddies.

Once they had reached their campsite and darkness had fallen, the two sons found a comfortable log to sit on and began flashing a pair of flashlights at the sky. Patricia's son Michael showed his parents that after pointing his light at a brighter-than-average star and making two short flashes, the two flashes were returned. John, the husband, ran back to the tent and found two larger, more powerful flashlights and gave one to Patricia. He next flashed two short flashes and one longer one. Within a few seconds, the very same sequence of flashes was returned.

Patricia flashed a different series which were also returned. Soon a few friends joined the single bright star. The whole family was having a great time, flashing away and getting their answers back, just as they had sent them. The couple theorized that the light show had been carried out by helicopters from a nearby air force base, with the pilots simply having fun with ordinary citizens. But what the heck, they'd had as much fun as the kids had.

The following day was uneventful, spent fishing and cooking up the day's catch. John planned to hike downstream in the morning to a place he had been told offered some of the biggest fish ever caught in the region. He checked the fire before the family retired. It was well stocked with wood and should still be burning when they arose in the morning.

Suddenly, about 1 A.M., Patricia was awakened by her terrified husband, who instructed her to get up immediately and put the kids in the car. Patricia was frightened herself, wondering what could possibly be so wrong that it would cause her husband to act like this. She and John packed the kids in the car and she pleaded to know what the problem was. His answer was a curt, "Never mind now! Just get in the car!"

A TERRIFYING PURSUIT

She did as she was told. John started the engine and they quickly moved down the dark road, leaving the campsite with all its elements intact. By now Patricia was becoming hysterical, still pleading with her husband to tell her what was happening. He turned to Patricia and said, "Look behind us! Do you see that bright light following us?" She looked out the rear window and replied in the affirmative.

The light was getting brighter and brighter, matching the speed of the car. John seemed to be like a lifeless zombie when Patricia yelled for him to stop the car. John quickly stopped, and the couple sat dumbfounded. They stared through the windshield in amazement. The car was now pointed in the opposite direction from the way they had just been heading. The road before them led back toward the campsite, not away from it.

When they returned to the campsite, Patricia noticed that the campfire had burned out. But John had set it to burn all night, and they had only been gone a few minutes. It was as if hours of time had just vanished. They decided together immediately that going home would be their next move. They were pursued by a greenish light that continued in its path for several miles. The couple felt a constant fear throughout the episode. The light disappeared after they turned onto the main highway. They were overjoyed to return to the safety of their home.

The next day Patricia was due for a checkup by her obstetrician, who told her that her pregnancy was proceeding normally and that she would have a healthy baby. Patricia has always been concerned about her daughter because of the strange experience on the camping trip. But her daughter Sonya was born on schedule and has grown into a marvelous type of human being with a high IQ and an uncanny sort of wisdom. Patricia also says that Sonya is much different than her other two children.

A TYPICAL BEDROOM ABDUCTION

In 1970, a year later, Patricia became aware of a strange greenish light bathing her bedroom as she and John slept. She felt she had lost her free will and could not command her own body. She soon found herself in a gigantic room sitting on a table with strange beings all around her. She could not consciously recall later what the beings looked like. Her eyes became fixed on three huge transparent cylinders, apparently composed of clear glass, which she estimated to be about 18 feet tall. Each cylinder contained a basketball-sized object similar to a ball bearing. This apparatus was making a sound, like the roar of the wind, which seemed to be coming from inside the cylinders. She noticed that she was on a platform of what appeared to be a balcony thirty feet above the cylinders. She thought to herself that, were she to fall, she would be killed instantly.

The next thing she recalled was being back in her bedroom lying next to her husband. When she yelled for him to wake up, he sat up straight with a horrified look on his face and asked what was going on and where the green light was coming from. Before Patricia could respond, the room was filled with a humming sound that seemed to be coming from outside. The greenish light began to fade and the room was again dark, with only the faint glow of the alarm clock remaining.

SURGERY NUMBER ONE

Patricia was the first person Leir and his team operated on. Leir provides a blow-by-blow account of the surgery as he worked alongside Dr. A. and Leir's surgical nurse, Denise. Every incision and soaking up of the blood with a surgical sponge is included in Leir's account, which adds to the credibility of the story in medical terms.

"All at once," Leir writes, "the sound of a crisp metallic click was heard. It broke the dead silence and reverberated in the small room. I had touched something with my probe and it had made a noise. I carefully inserted the clamp into the wound, spread the jaws to their maximum, reached in and grasped a solid object. I began the careful dissection to free the foreign body from its fatty and fibrous tissue attachments. In just a few moments I announced 'Here it comes!' and, with one final tug, the object came free from the inner confines of the toe. There was a sudden hustle and bustle in the operating room as we transferred the small object to a white sterile gauze sponge. All eyes were fixed on the peculiar object. Dr. A. peered intently at it and then glanced at me with a quizzical look. 'What in the hell is that?' he quietly exclaimed."

Leir found himself staring at what appeared to be a T-shaped or triangular-shaped mass that was dark gray in color and slightly shiny. It looked fleshy, not metallic, and was about one-half a centimeter long by one-half of a centimeter wide. The still photographer started to photograph the gauze sponge and its strange contents, followed by the videographer.

IMPOSSIBLE TO CUT

Dr. A. took the initiative and suggested that the team find out what was inside the exotic gray cocoon. Leir clamped the object with a heavy-duty surgical clamp and began to gently incise the biological covering. He was amazed to see that he couldn't cut through it at all. He asked for a fresh blade and tried a second time, again to no avail. He passed the object over to Dr. A., who also had no success cutting through the gray membrane. Dr. A. suggested they put the object aside and try again later. The object was then placed in the serum solution made from Patricia's blood for transport to a testing facility.

A second procedure was performed on Patricia and this time the implant proved to be a little harder to locate and remove, but Dr. A. stepped in and was able to cut the object loose. The object was placed on a clean sponge and then measured. The object resembled a cantaloupe seed and was also covered in the same smooth, glistening, dark gray covering as the first object. And, like the first object, the covering proved to be impossible to pierce with a surgical scalpel. The wound was stitched closed and Patricia was taken to a recovery room.

THE AMAZING GRAY COVERING

The first samples taken from Patricia and the other patients were the soft tissues that surrounded the objects, all of which

– amazingly – contained none of the inflammation that would normally surround a foreign body inserted into the flesh of a person. Leir told me in an interview that If we, as humans, could duplicate the gray membrane surrounding the recovered implants, we could wrap a heart or a liver in it and completely avoid the postoperative rejection by the body of a transplanted organ. The aliens seemed to have found a method for making the implanted objects meld with the human body in a way the body accepted without inflammatory "protest."

Leir was so surprised by this finding that he actually wondered if the human body had somehow undergone a fundamental physical change.

"All in all," he writes, "I found no medical literature that entertains the possibility that a foreign substance can be lodged in a human body without the surrounding flesh reacting to it. Our system of defense, called the reticuloendothelial system, comes into play the moment a foreign object enters our body. This system of defense is designed to ward off any invading substance, thus providing the body with the protection it needs.

"The transplantation of donor organs," Leir continues, "is a prime example of how our body rejects even the smallest differences in human tissue. If it were not for drugs that suppress the effects of our immune system, no transplant would be possible. The question for me was: Why did the body not produce an inflammatory reaction against the foreign bodies that we removed? A possible answer to this question became apparent after I had acquired the results of the next set of specimens analyzed."

SHADES OF GREEN AND PINK

Sims took the foreign objects Leir and his team had removed back to Houston. Sims worked with a chemist there who subjected the specimens to ultraviolet back light and found they fluoresced a brilliant green color. Sims was amazed because of the team's previous finding that a similar-colored fluorescence was present on the skin of a small percentage of abductees.

In another case, a pink stain was observed by both Leir and Sims on the hands of an abductee they examined who was subjected to the same black light phenomenon. Leir was able to remove the pink stain by scrubbing it with alcohol and its disappearance was confirmed with the ultraviolet light. It was concluded that the coloration must have been on the superficial skin only and not "subdermal." However, when the woman's hands were placed under the black light a second time a few hours later, the glowing pink stain had returned. Sims then smiled and told Leir that he had been right about the stain being subdermal after all. Leir had wiped it off the superficial skin, but after some time had passed, the pink hue had seeped back through to the surface again.

Back in Texas, Sims and his chemist associate dried the specimens, which made the membrane surrounding the objects brittle and therefore possible to scrape away from the metal object it surrounded. The membrane was found to consist of a protein coagulum, a substance derived from clotted blood, as well as hemosiderin, which binds with oxygen in a manner similar to its cousin, hemoglobin. The membrane also contained keratin, which comprises the outermost covering of our bodies. The combination of the various elements are not found in any tissues known to our current medical science.

The metal objects contained inside the biological membrane were subjected to metallurgical analysis using a method called a Laser-Induced Breakdown Spectroscopy (LIBS) at the Los Alamos National Laboratories. Some of the samples appeared to be metallic and contained aluminum, calcium, copper, iron, magnesium, sodium, manganese, nickel, lead, silicon, tin and zinc. A second round of tests, this time performed by the laboratory at New Mexico Tech, led the prestigious lab to declare that the metal samples material had most likely come from meteorites or fragments thereof.

WHAT ARE THE IMPLANTS FOR?

At this point, according to Leir, it would seem that the recovered items are structured objects that serve an as-yet-undetermined purpose.

"We hope that further study will provide answers regarding function," he writes. "I feel it is safe to put forth theories, but they must be looked at scientifically and either proven or disproven. One such theory pertains to the objects' abilities to function as tracking devices or transponders, thus enabling the subject to be found anywhere on the globe. Another possibility is that they may act as behavior-controlling devices. We know that abductees seem to have compulsive behaviors. I believe a more plausible purpose might be that they monitor certain pollution levels or even genetic changes in the body, which may be similar to the way in which we monitor our astronauts when they are in space. Only more time, effort and study will answer these questions."

ALIEN ARTIFACTS

Dr. Leir's work is very important on many levels. First of all, an abductee claims to have been abducted and implanted with some kind of foreign object. Secondly, the physical presence of the object is confirmed by x-rays to be in the exact place that the abductee claims it will be found. And finally, the object is surgically removed and manifests many strange properties, to include an incomprehensible lack of surrounding inflammatory tissue. In terms of investigating the physical evidence left behind in the wake of an alien encounter, nothing else even approaches the work of Leir and his team. We can hold the truth in our hands, but it remains much more difficult to understand just what that truth consists of.

ts Reserved   *Santa Cruz Herald*   VOL. CCXXXVI   May 12, 2006   NO. 102   WE/YF   * * *   *Fifty*

Alien object embedded in spinal column
Mystery Object Removed in Two Hour Procedure
Doctors Baffled; "We Have No Idea What this is."

By DS Goodman
AP Reporter

Santa Cruz—
An Aptos man says he suffered from back pain for years. Sometimes he had trouble walking. Sometimes he missed days at work. And once in awhile, he couldn't even get out of bed.

"The doctors said this, then they said that," complained Dan White. "They tried muscle relaxers, heat therapy, they made me exercise and stretch. Nothing worked. I figured I was just one of those people who had chronic back pain and would always have it."

But one day Whites luck changed. After appealing to nearly a dozen doctors over the years, White ended up the victim of a hit and run fender bender. He had a stiff neck, and went to the doctor for X-rays. Those X-rays revealed something more serious than a pulled muscle.

The two ounce "device" was removed from White's spinal cord without incident.

Doctor Hajduk of Dominican Hospital said, "We have no clue what this object is. We don't dare even speculate. Whatever it is seems to occasionally emit an ultra high frequency energy wave thats pitched far above the range of human hearing. I need to study this object further as this is the most interesting medical phenomenon I have come across in the 22 years of my medical career."

See "The Aliens Among Us Page D-5

Article from the May 12, 2006 edition, of the "*Santa Cruz Herald*" detailing a bizarre two ounce "alien implant" removed from the spinal column of a man named Dan White.

18.

TUMORS AND IMPLANTS
From the files of Lon Strickler
www.phantomsandmonsters.com/

We can go one giant step beyond alien implants - strange alien tumors as reported on the *"Phantoms and Monsters"* site by Lon Strickler. Something right out of the movie *"The Tingler"* with our old friend Vincent Price.

• • •

Over the years, I have received reports of alien implanted objects, seeded tumors and the transmission of chronic and terminal diseases. As well, other experiencers were cured of sickness and afflictions by alien beings...including cancer, crippling rheumatoid arthritis and birth defects. So...what is the real alien agenda? Is it simply non-discernible experimentation that results in good and bad effects? I am offering a few interesting emails I've collected from readers and others seeking answers:

Hi - This story involves an unidentifiable metal implant found in the cheek of a family member. This was found at

Parkland Hospital in Dallas, Texas, by X-Ray and MRI. This thing, whatever it is, was not something that we were ever previously told about nor was it anything associated with why he was receiving care at Parkland. They noted it several times and did not make any conclusions. Parkland made no mention of this thing to us or him. I discovered the reports of the implanted object after getting his medical files from Parkland from a private physician. The reports simply state there is a "metallic foreign body present in the right cheek." Oddly, the VA hospital in Dallas made no note whatsoever of the object in their MRI or X-Ray reports, yet there are numerous reports in the Parkland files. There is no wound accompanying the metal object. Our relative cannot feel it and did not know it was there.

This was discovered in May 2010 after he fell at home and the EMTs took him to Parkland, where extensive testing was done. Then, approximately four days later, he was transferred by ambulance to the VA hospital in Dallas.

He is retired and, over the years, I have wondered many times if the military implanted something in his body. He never was in battle or explosions but his job in the military was classified. I do know families of military guys who have said civilian hospitals found strange metal implants in such places as the mastoid area of their loved ones.

We really don't know what to think about this or the UFO issues. We just want to know what this thing is and if anyone out there has any information on these implants. At this time we have no idea if it should be included in any surgical removal plan or not. He is old and physically frail now, but his mind is clear as a bell, and of course we are wondering if that thing showed up in the films years ago and nobody said anything. I'd like to pull all those records. All of the other metallic items noted on the films, such as clamps and wires from heart surgery, were correctly and readily identified. Not that thing in his cheek. Please contact me if you have any info. I'd appreciate it very much. Thank you.

• • •

STRANGE ALIEN TUMOR NAMED "GILL"

Sacramento, CBS Local - When Josh Abken told his doctor he was having persistent back pain about a year ago, the doctor took an x-ray.

The Chico-area school vice principal said he was stunned to learn what was growing inside him.

"You think you got a serious muscle pull or something, [but] the doctor has that serious doctor face you see on TV," Abken said. "They didn't really know what was growing inside of me."

They named it Gill: A soccer-ball size "alien" tumor that had become as solid as a rock and even had growing tentacles.

Abken and his family made shirts bearing the motto "Kill Gill," based on the movie posters for the film "Kill Bill."

The husband and father of two said doctors told him the giant tumor was beginning to put pressure on his heart, shove his lungs out of position and even squash his stomach downward. It apparently had been growing for at least a decade.

"My kids are pretty young, but they knew Daddy had a big 'owee,'" he said.

Thoracic surgeon Costanzo Di Perna was able to remove the tumor before it killed Abken, but required several hours of surgery before it was all gone.

"Trying to take out this large, calcified tumor, hard like a rock, created a difficult surgical experience," Dr. Di Perna said.

Abken will continue to receive PET-CT scans to see if the tumor will return, but he has stayed cancer-free for about a year.

• • •

ANOTHER MIND SWIRLING CASE

Here is an interesting email I received from a reader in the UK back in January 2009:

"On January 22nd I had a major operation. They opened a 13-centimeter section of my spinal cord and removed a tumor. The doctors say they have never, in thousands of operations, ever seen a tumor like it.

"It looks like a salamander and has clearly defined eyes and a tail. I took a picture of it with my cell phone camera.

"It may be what was responsible for the weird visions I had of the nature of the universe. Since it was on my spinal cord it was directly attached to my brain.

"As I said before, it has what appears to be eyes and a mouth and the doctors reported some hard matter (spine?) in the tail of the creature. It was 8.3 cm long.

"I have asked the doctors for hi rez pictures. It has been sent to a lab for testing but results are not expected for 10 days. However, one of the doctors told me the tail seemed to have a bony structure in it. That is not something you would expect to find in a tumor made of nerve cells. Also, you can see what look like eyes. This really is bizarre.

"By the way, I feel just fine except for the fact they sawed open my back-bone."

• • •

I am writing to tell you of what my wife has gone through. She had a tumor between her shoulder blades that she could feel moving under her skin. When the tumor was removed it weighed 30 pounds and had tentacles and was moving on its own. The doctor said it was a lymphoma tumor. The only other thing that she was told is that they shipped it to the Mayo Clinic. My wife has had abduction experiences in her past. I am almost 100% certain that the so called tumor that she had was a direct result of those experiences. Thank you very much for the informative post as it has showed the both of us that she is not alone.

Sincerely, EF

• • •

Hi there...

I wanted to write and share one of my experiences and perhaps see if anyone knew of anything like it.

I've had so many similar experiences it's difficult to relate them. Especially since, many times, I just get "weirded out" and tired of trying to figure it out, or embarrassed by speaking about something that's happened that no one understands and so I deliberately force myself not to think about it. So here goes!

I had an experience many years ago in broad daylight with my then sister-in-law and many neighbors where we all watched an extremely large UFO for over an hour as it passed over from north to south, from the mountains north of San Bernardino, CA, over Riverside, and continued south. We were sitting on my apartment balcony waiting for our husbands to come home from work one summer day and noticed a strange looking dark cloud in the distance. The sky was clear except for this one huge, dark cloud. We watched as it moved fairly quickly from far to near and tried to figure out why it was there. The appearance, sound, altitude, speed, and direction of movement were all anomalous. It was black as a thundercloud and flashed occasionally but as it got closer, it was clearly lit from within with a strange pulsing yellow orange light and bright flashes, like lightning, but there was no thunder afterwards. It only "hummed" loudly with a very deep resonance. It looked almost identical to a huge storm cloud but at the same time not "normal." From my experience of watching aircraft and falling bodies all day in my skydiving days, I would say it was pretty low in the sky, maybe 2000 feet at most. It was surrounded by vapor that roiled around the body and drifted at the edges but never dissipated like a real cloud. The appearance was like an imitation of a cloud, with the exception of the low, loud hum, and didn't fit with the weather which was dry, hot, and with just the beginning of the usual evening breeze that blew in from the coast, west to east. We talked and watched the thing along with our neighbors in the complex who stopped to watch also. Some people were pulling into the parking lot from work and as they noticed us staring, they would join us and discuss it too before going about their business. We turned on the news to see if anything was being mentioned since this object was probably the size of a small block and so many were watching it. Nothing ever was reported. It was so strange but at the same time no one seemed at all troubled by it! We mentioned as we discussed it that even if we reported it, we'd never find out what it was and it would be pointless so why bother!

I sometimes think about it though and wonder what it was. Maybe someone else has seen it also or something like it. The year was between 1987 and 1989 because of where I lived, I'm thinking probably 1987. I was in Riverside, CA, at the time. Because it was coming from the high desert area I just assumed it was something the military was doing but, man, it was just gigantic and why would it look like a cloud? One thing I always thought about too is the huge number of people I knew who were present at the time who became seriously ill shortly after, including myself and my sister-in-law. She developed a form of thyroid disease that I can't recall now but it caused her eyes to bulge terribly and I have Hashimoto's thyroiditis. My youngest daughter, who was at home and present, has thyroid disease also, while my oldest, who wasn't in the area, is not affected. Several children in the area developed rheumatoid arthritis or other serious illnesses, like seizure disorders, also. Weird.

Regards, LAF

• • •

Hello,

Four years ago I experienced a horrific alien abduction that lasted 3 days. Several weeks after the incident I noticed that I had a very stiff neck when I would wake in the morning. Eventually, the pain increased to the point where it bothered me all the time. I went to my physician and had X-rays and a CAT scan where it was determined that a foreign object that was made of a metallic material was embedded in vertebrae very near my spinal cord

To make a long story short, I had surgery to remove the object. The surgery was performed at the Long Beach VA Hospital in California. I had requested that the object be preserved because I wanted to have it analyzed. I was told that the object was "lost" after surgery but I was later told by an attending nurse that it was immediately shipped off to an Army research facility. She said that she got a look at the object, which was about a half-inch in length and was made of a black colored metal. She also said that the object was difficult to capture during surgery and seemed to "move' away from the surgical instruments.

I am interested in other experiences and thoughts. I would like some answers. Thank you, CF

ALIEN ARTIFACTS

BOOKS BY LON STRICKLER AVAILABLE ON AMAZON

THE MEME HUMANOIDS

ALIEN DISCLOSURE: EXPERIENCERS EXPOSE REALITY

CRYPTID ENCOUNTERS

STRANGE ENCOUNTERS

ADDITIONAL READING

Casebook: Alien Implants
(Whitley Strieber's Hidden Agendas)

***The Alien Abduction Files: The Most Startling Cases
of Human Alien Contact Ever Reported***

***The Aliens and the Scalpel: Scientific Proof of
Extraterrestrial Implants in Humans***
(New Millennium Library, V. 6)

***Alien Nightmares: Screen Memories of UFO Alien
Abductions: Abducted by Aliens for Decades***

UFO-related material from the Advanced Aerospace
Threat Identification Program show that encounters
with UFOs have reportedly caused radiation burns
along with brain and nervous system damage.

19.

CAUTION: ARE UFOs A CONTAGIOUS DISEASE, TOO?

By Gene Steinberg

I've followed our paranormal world for an awfully long time. Over the years, I've made a big point of the fact that UFO sightings do not always exist alone. People may see one and get on with their lives, while others find themselves confronted with all sorts of incredible phenomena, sometimes extremely frightening, quite possibly painful.

In some cases, there's a family history of encounters with the paranormal, but in others there isn't. It all begins with that one sighting and then everything goes to hell.

But when family is involved, quite often the tales of what went beforearen't always revealed to a younger family member, unless they have a need to know. And that need is their own paranormal experience.

TOO UPSETTING TO DISCUSS

As regular readers of my articles and listeners to my radio show, "The Paracast," know, I've written about this before. People I know have had their lives turned upside down as a result of such experiences.

I recall an incident in the late 1960s, when I was working at a small radio station in the Midwest. One of my friends from the UFO field lived nearby, so I planned a visit.

But when I arrived, he told a terrifying tale involving him and his then-girlfriend, both of whom saw a UFO. It was followed up by frightening events in his home, including an attack on the woman by an invisible entity.

I have not followed up, because he clearly indicated it wasn't something he'd care to talk about anymore.

HEALTH PROBLEMS THAT GO ALONG FOR THE RIDE

In addition to such otherworldly phenomena, there are reports that some UFO encounters are not quite positive for your health.

So when he appeared on "The Paracast" back on November 28, 2021, Colm Keller, a coauthor of "*Skinwalkers at the Pentagon: An Insider's Account of the Secret Government UFO Program*," he spoke of a "hitchhiker effect." It means that, when some people see a UFO, it's not just about watching something strange. It means that "something" starts tagging along as you go on about your business.

On the surface, it almost seems as if you're catching a disease. And in some cases, that appears to be true. Suddenly unexpected things occur, such as the onset of possible poltergeist phenomena. Things are suddenly moved around the room without visible cause, and I think of the early scenes of the 1982 movie, "*Poltergeist*," for some visual, if exaggerated, examples.

But when it comes to catching a disease, perhaps the reality is more frightening than having a few plates flip around the kitchen of their own accord.

The 1977 movie, *"Close Encounters of the Third Kind"* featured scenes where one of the protagonists, Roy Neary (Richard Dreyfuss) suffers from extreme sunburn after a UFO sighting. It's not an uncommon phenomenon. But the subsequent scenes, where he and a small number of other eyewitnesses are drawn to the site of an upcoming UFO landing, are best left on the cutting room floor for this discussion.

MULTIPLE SYMPTOMS AND LONG-TERM EFFECTS

Unfortunately, there are UFO cases where witnesses experience sometimes painful health effects. One well-known and key example is the December 29, 1980, Cash-Landrum encounter, where Betty Cash, Vickie Landrum and the latter's grandson, Colby Landrum, then aged 7, had a close-up sighting of a large diamond-shaped object hovering at treetop level in the state of Texas.

It's notable that the object reportedly emitted flame and heat, according to reports on the sighting. The physical effects may explain the sort of nasty health consequences that occurred. Indeed, the heat impacted the car's body to the point where the exterior became painful to touch. Whatever the cause, all suffered health effects over the years, with Cash evidently experiencing the worst.

They all evidently suffered from nausea, vomiting, diarrhea, feelings of weakness and burning sensations in their eyes soon after the encounter. It's as if they had sustained a massive sunburn, or were they irradiated by the heat-emitting flying triangle?

What about possible radiation exposure?

Over the years, Cash developed breast cancer. The elder Landrum had cataracts. There has been speculation about what sort of emissions would cause such side-effects.

Except for the initial onset of symptoms, however, it's hard to correlate what happened with the later outcomes, obviously. During their lifetimes, one in eight women, based on U.S. estimates, will suffer from breast cancer.

**Vickie Landrum, Colby Landrum
and Betty Cash.**

JUST A PART OF LIFE

Cataracts are even more common, with one in six Americans over the age of 40 developing them. The average age for cataract surgery is 73. My two cataract surgeries occurred just before I reached 76, and I have certainly not had any close encounters with a UFO, or none that I know of. It's just how things are.

Indeed cataract surgery appears to be so normal that the eye surgery center that I used runs the procedures in supermarket fashion. After prep time with eye drops and light sedation, the actual surgery to replace the lens with an implant typically takes about three or four minutes, roughly speaking. My doctor did the narration in proper sportscaster fashion, and, when done, he quickly moved on to the next patient.

It's quite normal, obviously, that medical needs increase as one gets older — and are just a part of life. That said, being irradiated by a UFO and suffering health effects after the incident raises questions as to just what is going on here.

Unlike most UFOs, the craft seen in the Cash-Landrum case appeared less advanced. It expelled flame and heat, unlike the flying things that appear to possess no visible propulsion system.

Indeed, it is not untoward to wonder if this particular UFO might have been some sort of test aircraft that flew about without the expectation that nearby eyewitnesses would suffer physical effects. Or maybe the authorities didn't care.

THE RENDLESHAM FOREST INVESTIGATION
AND ITS AFTERMATH

Another 1980 sighting, at Rendlesham Forest in the UK, also involved a witness suffering from possible physical effects. Retired technical sergeant John Burroughs and fellow airman Jim Penniston were the first to investigate reports of mysterious lights located near the East Gate of RAF Woodbridge.

It was 3:00 AM on December 26, 1980, three days before the Cash-Landrum sighting. Lights were observed descending into nearby Rendlesham Forest. But as servicemen approached the phenomenon, they saw what was described as a glowing, possibly metallic object with colored lights.

That morning, the serviceman found three small triangular impressions on the ground, along with burn marks and broken branches. The physical evidence was observed near the eastern edge of the forest at a small clearing.

Two days later, as servicemen were investigating the site; a flashing light was seen across the field to the east, almost in line with a farmhouse, similar to what the witnesses reported on the first night. The brightest of the three lights hovered for two to three hours, occasionally emitting a beam of light.

In this case, authorities suggested that they were caused by stars.

While the Rendlesham case was popularized over the years, not all UFO researchers believed it. Jenny Randles, co-author of an early book on the topic, *"Sky Crash: A Cosmic Conspiracy,"* evidently had a change of heart.

Technical Sergeant John Burroughs

In 2010, she was quoted as saying, "Whilst some puzzles remain, we can probably say that no unearthly craft were seen in Rendlesham Forest. We can also argue with confidence that the main focus of the events was a series of misperceptions of everyday things encountered in less than everyday circumstances."

Another prominent British UFO researcher, David Clark, announced in 2018 that the incident was a set-up or hoax perpetrated by the UK's Special Air Service on the U.S. Air Force.

Yet another author has had second thoughts. Peter Robbins, who co-wrote *"Left at East Gate: A First-Hand*

Account of the Bentwaters/Woodbridge UFO Incident," disavowed the book after discovering that his co-author, Larry Warren, had been making up tall tales about his experiences.

STRUGGLING WITH THE VA OVER HEALTHCARE COSTS

But that didn't stop Burroughs, who continued to suffer physically after the sighting. His ailments included heart problems, and, as a result, he now has a pacemaker.

After struggling to collect his rising medical expenses from the VA, he finally won a settlement.

The resolution of the case came over three decades after the sighting, after he endured years of continuing denials of coverage from the VA. We stayed on top of it, so on the March 1, 2015, episode of "The Paracast," we featured Burroughs, along with former UK Ministry of Defence official Nick Pope and attorney J. Patrick Fascogna, who negotiated the settlement.

A key piece of evidence that helped resolve the dispute was a formerly classified study named Project Condign, which reportedly revealed critical details of the Rendlesham events, concluding that it "might be postulated that several observers were probably exposed to Unidentified Aerial Phenomena (UAP) radiation."

Now, in light of the claims that the Rendlesham case resulted from a misidentification of conventional objects, or was, in part, the result of a hoax, the conclusion in the Condign report appears especially curious. Wrong or right, we may never know. Burroughs nonetheless had his treatments covered.

DIFFERING FORMS OF CONTACT APPARENTLY LINKED

Regardless of the actual cause, the Cash-Landrum and Rendlesham cases are among the exceptions. Aside from the sunburn effect, most people don't appear to suffer any ill effects from a UFO encounter. Where otherworldly things occur, it's about external phenomena, possible poltergeists in one's home or the appearance of strange creatures in the neighborhood.

So when investigators look into reports of possible ghosts and strange creatures, particularly the latter, it's not unusual to find UFOs as part of the picture.

But which is the chicken or the egg? What begat what? Was it the UFO, causing someone to suffer from the disease of psychic sensitivity, thus making them more likely to witness things they wouldn't have normally seen? Possibly, but how does that cause physical effects, too?

Unless they are infected with the ability to generate such effects in their surroundings.

Unfortunately, many UFO researchers aren't in tune with potential human effects, whether illness or the onset of other paranormal experiences. They have decided that, when you see a UFO, you might as well be watching an airplane going about its business. There is no difference, and researchers who consider a wider picture are engaged in pursuits that just aren't scientific.

CAN WE EXPECT MORE OPENNESS?

But things may be due for a change.

The new UAP office that was established in the 2022 U.S. National Defense Authorization Act (NDAA), according to a press release from Senator Marco Rubio (R-FL), one of the sponsors of the amendment, "will be required to provide unclassified annual reports to Congress and classified semiannual briefings on intelligence analysis, reported incidents, ***health-related effects*** [my emphasis], the role of foreign governments, and nuclear security."

Those three words, "health-related effects," have gone by without much coverage. But it also appears to be a key component of the new investigation. It might have even been due to the influence of Dr. Kelleher and the others who were engaged in that infamous $22 million Pentagon UAP study.

Unfortunately, it doesn't appear as if the mainstream media will pay much attention to what it all means. That there is a UFO/UAP study is the story in and of itself. Whether they will consider whether ET is among us will surely get its share of coverage, although it's being done in a way that deemphasizes the possibility. Instead, it's about the potential impact to national security.

It also gives them an out should someone ask Pentagon officials about what "health-related effects" truly means.

But if there is cause to pursue that avenue of research in connection with any sighting that becomes public, the authorities can always maintain that, well, of course, they have considered all possibilities.

Of course, old-fashioned UFO groups, such as MUFON, aren't hard-wired to expand the scope of their investigations into such side issues, which is unfortunate. As the result, key evidence to decode the UFO puzzle is probably being ignored.

The bigger question is whether UFOs are part of a larger range of paranormal phenomena and aren't separate after all. A proper investigation ought to consider all logical — and maybe illogical — possibilities and not focus on single, simple solutions.

But it's fair to say there is little hope that it'll all be resolved soon, if ever.

• • •

Veteran UFO researcher Gene Steinberg is Host and Executive Producer of "The Paracast" (www.theparacast.com), considered since its 2006 debut to be "the gold standard of paranormal radio." In his "other life," he is the author of numerous articles and over 30 books on personal technology, plus two sci-fi novels.

Rumors have circulated for years that the U.S. has managed to capture a number of crashed UFOs, and that these UFOs have exposed people to unknown, possibly alien, pathogens.

20.

THE POSSIBLE BIOHAZARD THREAT OF UFOS
By Tim R. Swartz

This chapter will deal with the "extraterrestrial hypothesis" of UFOs and the possibility that if alien life-forms are visiting Earth, they may also be bringing their viruses and bacteria along for the ride. Of course I have my doubts that the UFO phenomenon can be bookmarked solely in the nuts and bolts "visitors from other planets" theory. But, whatever their origins, there have been some interesting cases that show a disturbing relationship with suspicious illnesses and deaths to witnesses who claim to have had a close contact with UFO-related creatures.

Seeing how vulnerable humans can be to the microorganisms that are indigenous to our planet, exposure to something completely new, to which we have no natural immunity, is a frightening thought. Considering that, what are the chances that viruses and bacteria from other worlds would even find our DNA palatable?

WE COULD ALL BE RELATED

Panspermia...a theory that suggests that life on Earth originated with help from microorganisms and biological material from outer space. It's an intriguing idea, that life here got its start from somewhere out there. But, if that were the case, it probably wasn't a "one off" situation. Space could be filled with extraterrestrial bacteria and viruses, travelling on cosmic dust between the stars, seeking new homes to grow and flourish.

The late, great astronomer Fred Hoyle, along with Chandra Wickramasinghe, wrote a book about this in 1979 called *"Diseases from Outer Space — Our Cosmic Destiny,"* where they propose that many common diseases, such as influenza, the common cold and whooping cough, have their origins from outer space.

Wickramasinghe has even recently suggested that the coronavirus, which causes Covid-19, could have been brought to Earth by a piece of space rock during a brief fireball event over China in October 2019. Naturally, other scientists have refused to take Wickramasinghe's theory seriously, saying that his ideas are pseudoscience. There is plenty of evidence that the coronavirus, called SARS-CoV-2, lines up with other known terrestrial viruses that have caused similar outbreaks in the past. But we shouldn't dismiss Wickramasinghe with a quick admonishment of "bad science" simply because the idea that viruses and bacteria might be able to survive the harsh conditions of outer space is "outlandish." Life has the amazing ability to be incredibly tenacious and thrive under the most inhospitable of conditions.

For example, the Russian space station Mir encountered problems with a film growing over its windows. This hurt the crew's ability to see outside while in orbit. Moreover, the small organisms had done more than just coat the window and block astronauts' views of space...they had actually damaged the window which was made of quartz glass in a titanium frame encased in enamel.

Closer examination of the Mir space station revealed that the bacteria and fungi had also harmed electronic equipment by rusting copper cables and coating several other surfaces.

Scientists were so surprised because space vehicles are cleaned with toxic gases before being sent into space. How did these fungi and bacteria survive and thrive once in outer space? It is possible that any surviving organisms, in a sterile environment without other organisms around to compete, thrived and multiplied. Even more startling, cosmic radiation, instead of killing them, may have caused them to mutate and grow faster than they normally could on Earth.

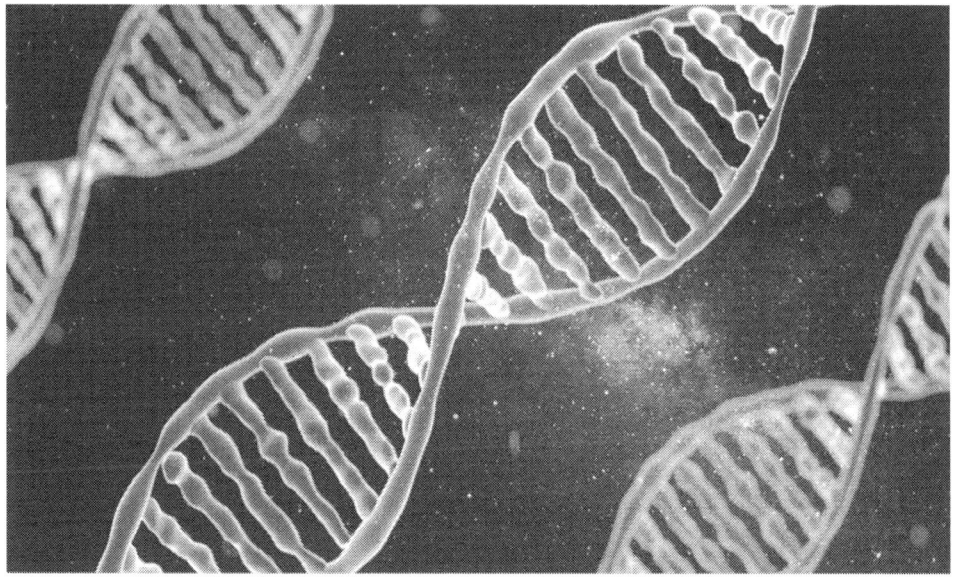

It may be possible that life permeates the universe and is able to travel between the stars by hitching a ride on comets, asteroids or even cosmic dust.

These findings were supported by experiments done on the International Space Station (ISS). In 2015 a box of exposed microbes were mounted outside of Japan's Kibo lab on an exterior handrail. The hearty bacteria, known as Deinococcus radiodurans, had no protection from an onslaught of cosmic ultraviolet, gamma, and x-rays. The final results showed that the Deinococcus bacteria survived the three-year experiment. The bacteria cells in the outer layers of the masses died, but those dead outer cells shielded those inside from irreparable DNA damage. And when the masses were large enough, still thinner than a millimeter, cells inside survived for several years.

One could not ask for a better "alien artifact" than microbial life raining down on us from the deepest regions of space. However, we will take it a step further with some cases that could show that some close encounters with UFO occupants may have resulted in exposure to deadly microbes that did not originate from this planet.

THE STRANGE DEATHS OF VARGINHA, BRAZIL

In January of 1996, a number of people around the town of Varginha, Brazil, claimed to have encountered one or more bizarre creatures, possibly associated with a crashed UFO spotted on January 13, 1996somewhere between Varginha and the neighboring city of Três Corações.

One week later, on January 20, three girls were coming down a narrow path in the area known as Jardim Andere, located a mile and a half away from the center of Varginha, when 16-year-old Liliane Fatima Silva spotted a strange creature, with oily-looking brown skin and what looked like three small rounded horns protruding from its head. It was near the back wall of an old garage.

"He was squatting next to the back wall of an old garage, with his long arms between his legs," the girl said. "His eyes were huge and red."

Liliane's sister, Valquiria, 14 years old, and their friend Katia Andrade Xavier, 22 years old, told investigators that the creature wasn't an animal and certainly wasn't human.

The girls also said the creature "gave off an unbearable smell of ammonia."

"It was a horrible thing." says Katia.

"He seemed stupefied. He didn't make any noise," says Valquiria. At that point, the creature moved as if to stand up and the three girls ran away.

Investigator Ubirajara Rodrigues, a lawyer and university professor, who had been researching UFOs since the late 1970s, reported that military police (the equivalent of state police in the U.S.) had captured a creature in woods three blocks away from the vacant lot. (Some witnesses say that firefighters caught the creature). One witness said that they saw armed soldiers searching the same woods, hearing three shots being fired, and then seeing two body bags (one squirming, one still) being loaded onto an army truck...possibly meaning there were two creatures involved.

Apparently, the creatures were taken to the Regional General Hospital of Varginha. The still living creature stayed there for a few hours and then both were transferred to a better equipped facility, the Humanitas Hospital. The dead creature was kept at the Humanities Hospital for two days when, on January 22, a huge military operation involving three Army trucks, each one driven by two different soldiers, took place to remove the body.

All three trucks then transferred the cargo to a military facility in Campinas, State of Sao Paulo, about 200 miles from Varginha. There, the corpse was removed to the University of Campinas. No additional information about the possible still living creature's fate has ever surfaced.

A few weeks later, on February 7, 23-year-old Corporal Marco Eli Chereze, one of the military police that captured, and touched with his bare hand, the living creature, underwent surgery to remove a small tumor, similar to a furuncle, from his right armpit. However, his condition quickly deteriorated and he started to complain of severe pain followed by paralysis. At this point he was transferred to the Regional Do Sul de Minas Hospital where he was admitted to the ICU. On February 15, he died of septic pulmonary thromboembolism.

Cardiologist Dr. Cesário Lincoln Furtado told investigators that another test detected an unexpected immunodeficiency in the patient.

"At the beginning, the diagnosis of a urinary or kidney infection prevailed because of the presence of 'enterobacteria.'" Dr. Furtado said. "But, in less than seven days, three kinds of bacteria attacked the policeman. This is a very rare thing.

"The bacteria detected in the exam were sensitive to antibiotics, but the patient didn't react. He was tested for HIV, but that came out negative. The patient's immune system had disappeared, but it was never known what caused it, as no virus or other cause was detected in the scans at the time.

"I have no idea what happened to this poor man," Furtado said. "It was a strange death, without rational explanation."

Another bizarre aspect of Corporal Chereze's death was that his wife was never given a death certificate and did not attend his funeral. This was because, as she said, "I was not told anything about the funeral. It was all secret and handled by the military."

UNKNOWN ILLNESSES

In early 2022, UFO researcher and reporter Edison Bonaventura interviewed Saulo José Machado, who claimed he was a low-ranking junior soldier involved with the capture of one of the Varginha creatures. The video of this interview was posted on Bonaventura's YouTube channel, Enigmas e Mistérios (Riddles and Mysteries).

Machado said that he, along with other soldiers from his base in Belo Horizonte in southeastern Brazil, was sent to Varginha for an undisclosed operation. While sweeping the area where they were dropped off, he claims two other soldiers came out of the woods carrying a strange being.

"The first thing that caught my attention was the very large head. The head was completely disproportionate to the body with very big red eyes. And then, in those 10 or so seconds that I had to observe it, I noticed that it was a dark, reddish-colored, and very oily looking. It also had a strong smell that reminded me of acetone."

Reconstruction of the strange creature seen by three girls in the city of Varginha, Brazil.

Machado also said that he, along his fellow soldiers that he managed to keep in touch with, all suffered from mysterious and debilitating illnesses that have so far mystified all doctors' diagnoses.

MYSTERY AT THE ZOO

The bizarre events at Varginha slowly began to die down, to the point where people began to feel that everything was now behind them. Unfortunately, there was still plenty of high strangeness to go around for the Brazilian town.

On April 21, Mrs. Terezinha Gall Clepf was attending an event at a small restaurant located in the Varginha Zoo. Around 9:00P.M., Mrs. Clepf was standing on the restaurant's porch smoking a cigarette when she spotted a strange creature no more than 13 feet away.

The women said the creature was standing behind the railing that surrounded the porch and was clearly staring at her. She described it as similar to what was described by the three girls and the military several months earlier. She said the eyes were enormous, red, and emitting a type of luminescence which allowed her to clearly see its face.

The main difference between what she saw and what the three girls described was that this creature had what appeared to be a golden-colored helmet on its head. Mrs. Clepf ran inside the restaurant to get her husband. Unfortunately, when the two came back out, the creature had vanished.

Whether or not there was any connection to what Mrs. Clepf saw at the Varginha Zoo, and what happened next, is open to speculation. But it does seem odd that after that point, animals began to mysteriously die at the zoo.

Biologist Leila Cabral, who was Director of the zoo at the time, remembers clearly the strange circumstances.

"The animals started dying in a very strange way," Cabral said."They just died. About five animals died within days, something that had never happened before at the zoo. There was no plausible explanation."

The dead animals consisted of two brocket deer, a tapir, a blue macaw and an ocelot.

After the deaths, Cabral and the zoo's veterinarian performed necropsies on the animals and sent samples for examination. A laboratory in Belo Horizonte (MG) detected an unidentified toxic-caustic substance in one of the deer tissues, but it wasn't detected in any other samples.

Veterinarian Dr. Marcos A. Carvalho Mina, who later took over as Director of the Varginha Zoo, confirmed the information and also adds that the necropsy detected a blackening of the animals' stomach and intestine mucosa. Poisoning was even considered, but as nothing was detected, that possibility was ruled out.

"All the animals had the same symptoms at necropsy, but they were animals of different species and they feed in different ways," Dr. Mina recalled. "They were in different places [at the zoo]. There were also no bacteria [in the exam], so that's what makes the situation complicated."

The exact reasons for the animal deaths could not be determined. Dr. Mina came to the conclusion that the entire episode was merely a coincidence. This view was not shared by Dr. Leila, who decided these deaths had something to do with the strange creature seen by Mrs. Clepf.

It has been suggested that the creatures captured in Varginha, whatever their origins, may carry bacteria or viruses that are capable of contaminating and killing both humans and animals just a few days after exposure. Could this be a tangible reason for the Brazilian military to hide the truth from the public? The answer to this question remains unknown.

RESEARCHER TALKS UFO BIOHAZARD ISSUES

For years, UFO investigators have used Geiger counters and EMF detectors to look for radiation and electromagnetic fields after UFO sightings. But, according to Donald Burleson, the New Mexico State Director of MUFON, there is a biological component to the UFO phenomena that many in his field overlook.

ALIEN ARTIFACTS

As reported by the July 3, 2021 edition of the *Roswell Daily Record*, (by Alex Ross), Burleson, in a lecture given at the International UFO Museum and Research Center in Roswell, New Mexico, told the crowd that biological pathogens carried by extraterrestrials resulted in the deaths of members of UFO recovery teams, including in the famous 1947 Roswell incident.

He said one example of this happened in the Roswell incident, when a rancher in July 1947 saw what some believe was a UFO that crashed.

Burleson said four technicians involved in the recovery of the downed craft died mysteriously soon after handling the craft. He said the four technicians abruptly experienced seizures, bleeding and other symptoms of hemorrhagic fever. All four, he said, were taken to the research laboratory in Los Alamos for observation and died.

"Somehow it is easy to assume that you can just walk up, grab the bodies and stick 'em in a bag or something, but I would be a little careful about doing it," Burleson said in reference to the retrieval of alien bodies from crashed UFOs.

Burleson said documents from MJ-12 — a government team assembled after the Roswell incident and referenced in UFO researcher Robert M. Wood's book *"Alien Viruses: Crashed UFOs, MJ-12 & Biowarfare"* — show that as early as the late 1940s scientists were likely performing research on the body tissue of aliens.

He said a report from MJ-12 released in 1951 and referenced in Wood's book talked about viruses and bacteria found in the bodies of aliens so lethal they could launch medical science into whole new fields of biology and possibly biological warfare.

Burleson also mentioned one alleged incident, in which he said a UFO crash very likely could have resulted in a deadly pandemic. He cites the book *"Mexico's Roswell: The Chihuahua UFO Crash"* by UFO researchers Noe Torres and Ruben Uriarte.

The book explores what is speculated to have been an Aug. 25, 1974 collision between a commercial airliner and a UFO just south of the U.S.-Mexico border that landed in the Mexican state of Chihuahua.

U.S authorities, Torres and Uriarte say, tried to access the craft and the airliner but Mexican officials declined to grant them permission. Instead, Mexican officials assembled a recovery team to retrieve the plane and the UFO, which were placed on flatbed trucks.

Burleson said Torres and Uriarte state the CIA then flew a clandestine mission in Mexican air space, with the intent to monitor the convoy, which was traveling south. However, CIA officials, when they encountered the convoy, found it had come to a halt with all members of the Mexican recovery team dead from what is believed to have either been a chemical leak or biological agent.

Burleson said he believes a U.S. Team took the UFO wreckage to Atlanta because that is where the Centers For Disease Control Level 4 Lab is located. He added that facility is one of the few places that had the capability to deal with a biohazard and keep it secret. He said the U.S. response was necessary and very well could have spared the world from a deadly pandemic.

DEADLY ENCOUNTERS

In his lecture Burleson cited MJ-12 documents as a source for possible government research into biological hazards associated with UFOs and their occupants. It cannot be stressed enough the possibility that the MJ-12 papers are fake, or at least part of an elaborate disinformation campaign.

The particular MJ-12 papers detailing the UFO-related biohazards were released by Dr. Robert Wood and Ryan Wood, the father-and-son team devoted to analyzing the MJ-12 documents that have been fed to UFO researchers since the early '80s.

The Interplanetary Phenomenon Unit Summary of the IPU contained information concerning two UFO crashes that occurred on July 3, 1947 outside of Roswell, New Mexico, and designated as LZ-1 and LZ-2.

The IPU Summary suggested "that at least some of the alien bodies found at UFO crash sites in New Mexico in the summer of 1947 presented a biological threat." Mentioned was an autopsy of an alien cadaver by Major Charles E. Rea and Dr. Detlen Bronk. Both doctors actually existed and did have impressive backgrounds. Also implicated was Dr. Stafford L. Warren of Oak Ridge Laboratories, as there was "an indication that Warren had an awareness of pre-1947 references to fatal encounters with an alien virus."

Majestic Twelve Projects: 1st Annual Report was prepared by a "Special Panel" linked to MJ-12. The report said in part: "The samples extracted from bodies found in New Mexico have yielded new strains of a metro-virus not totally understood, but give promise of the ultimate BW weapon.

"Autopsies on four dead SED technicians indicated that the four may have 'suffered from some form of toxin a highly contagious disease.' Those samples are kept at Fort Detrick, Md.

"...Personnel at Camp Detrick/Fort Detrick were concerned specifically about 'airborne' outbreaks of serious diseases in the exact year that the New Mexico UFO crash-retrievals occurred," said the authors.

The papers also indicate: "A crashed UFO near Kingman, Arizona, in May, 1953, suggested that the 'retrieval team' may have been exposed to the same deadly virus that broke out in a 'mysterious sickness' in 1947."

Another classified document, dated March 24, 1995, stated about the 1953 retrieval: "...the crew after one hour emerged from the craft confused, with upset stomachs, removed their masks and threw up."

A number of UFO crashes are mentioned, such as the 1964 Big Thicket, Texas crash, and others, in 1958 (Arizona) and Northern New Mexico, near Taos, in 1961: "....the retrieval teams were overcome with nausea and dizziness, and at least two died of a 'flesh-eating virus – kind of like Ebola'..."

In the Big Thicket case, the team from Fort Hood were "garbed in biohazard and protection suits." Will, a witness at Big Thicket, often wondered how many spheres may have landed, unknown to the military, and may "have possibly unleashed such viruses upon the local population."

If we are being visited by creatures from other worlds, be they extraterrestrial, ultraterrestrial or interdimensional, what are the dangers such visitors could produce for any poor human who may accidently stumble upon them? It is bad enough that we could be exposed to extreme radiation, electromagnetic fields or excessive heat. But think of the possible horrors of exposure to alien microorganisms to the human body with no natural immunity. That being said, caution should always be taken when dealing with UFOs, as history as shown that – in one way or another – UFOs are hazardous to your health.

ADDITIONAL READING

UFO CRASH IN BRAZIL – A GENUINE UFO CRASH WITH SURVIVING ETS

By Roger K. Leir

Rumors have circulated for years that the Earth has been infiltrated by extraterrestrials that, with the exception of certain features, appear human.

21.

ALIEN ARTIFACTS OF THE DNA KIND
By Tim R. Swartz

Skeptics of the UFO phenomenon repeatedly point out that there is no tangible evidence to support the claims made by both witnesses and researchers. For the most part, evidence for the reality of the UFO phenomena consists of fragments of metal, some mysterious fluids, and the occasional pancake or two. However, there is a case of mysterious physical evidence that goes way beyond chunks of metal, and into the very framework of life itself. This is the bizarre event of July 23, 1992 involving Peter Khoury of Sydney, Australia, who found himself in possession of a rather intimate alien artifact.

Peter Khoury was no stranger to the strange world of UFO abductions. Khoury was originally from Lebanon and moved to Australia in 1973. On July 12, 1988, Khoury was lying in bed when he suddenly found himself paralyzed and surrounded by three or four figures wearing dark robes with hoods on their heads. He was told telepathically to relax and that he would not be hurt. Two of the figures came close enough that Khoury could see that they were a golden-yellow in color, tall and thin, with big black eyes and a narrow chin. The one closest to him brought out a long needle-like, flexible crystal tube which was inserted into the side of his head, causing him to lose consciousness.

When he awoke Khoury found his father and brother asleep in front of the TV. When he woke them, they were dazed and believed that only 10 minutes had passed, but it was actually more than two hours. At that point Khoury was not familiar with UFO abduction cases and he battled with anxiety and confusion because of his strange experience.

BEDROOM INTRUDERS

By 1992 Peter had married and was living in Sydney. On July 23, he stayed home for the day because of a work-related injury. After driving his wife to the train station, he returned and went back to bed. Khoury would later tell UFO researcher Bill Chalker that around 7:30 AM he awoke suddenly, realizing that he was no longer alone in the house.

Peter Khoury and Bill Chalker look through the book "*Hair of the Alien*," which recounts Khoury's bizarre bedroom encounter.

"I was trying to wake up, put my senses together," Khoury said. "I noticed there were two naked females on the bed. The one directly opposite me was a blonde-looking woman and she was sitting with her legs tucked under her. The one on the other side was dark-haired and oriental-looking. She was kneeling on my bed, kind of sitting upright a bit."

Khoury said that the blonde woman appeared to be in her mid-thirties with white, almost translucent skin. Her hair was thin and wispy, but was curled "something like Farrah Fawcett, really exotic." The shape of the face was long, like it was stretched out with an extremely pointed chin. She had high, protruding cheeks, with a long nose. Khoury also noticed that her eyes were two to three times bigger than normal eyes.

"She looked humanoid," Khoury said. "But I knew I wasn't looking at a human female."

The other woman, Khoury remembered, seemed to be around 5 foot 8 inches tall, but her features weren't completely human either. Her skin was dark, like an Indian. She had straight, stiff black hair that went down to her shoulders. Her cheekbones looked Asian, but they sat up too high on her face and were puffy, like she had been "punched by Mike Tyson."

"Her eyes were large and dark, almost black. I don't remember seeing white in the eyes."

"THIS ISN'T THE WAY IT'S SUPPOSED TO HAPPEN"

The blonde seemed to be in charge and Khoury had the impression that she was communicating telepathically with the dark-haired woman. The blonde woman then reached forward, cupping her hands around the back his head, and pulled him to her breast. He would pull away, but the woman would pull back even harder, trying to force her nipple into his mouth.

Khoury told Chalker that, in his panic, he bit down hard on the woman's nipple, accidently swallowing a piece of it. The bizarre thing he said was that the blonde woman did not cry out or act like she was injured. Nor was there any blood. Instead, the expression on her face was like shock or confusion.

"She looked at the Asian one," he said, "and looked at me like, 'This isn't the way it's supposed to happen. You've done this wrong.'"

Involuntarily, Khoury swallowed the nipple and it caught in his throat, causing him to have a coughing fit. When he looked up, the two women had vanished.

Continuing to cough, Khoury went to the bathroom for a drink of water, but realized that something was hurting his penis. When he checked, he found to his dismay that there were two long strands of thin, blonde hair wrapped tightly around his penis underneath the foreskin.

As the pain became an intense burning sensation, he finally managed to remove the hairs and immediately put them in a small sealable plastic bag.

"The reason I did that was because I knew that there was no way, no way at all, that a hair that size and wrapped around the way it was should have been there," Khoury said. "Thinking of these women, the thing in my throat, the hair, something bizarre had just happened."

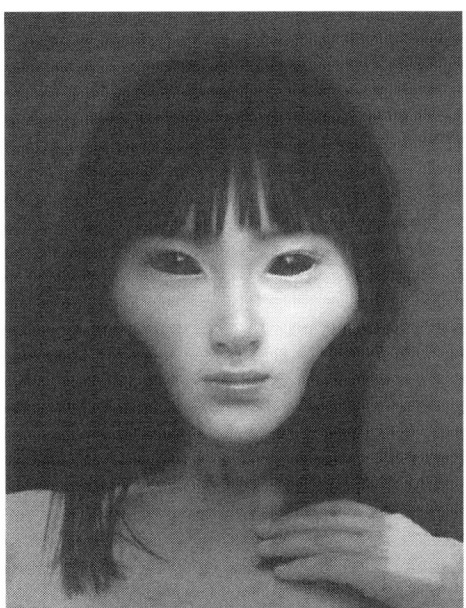

One of Khoury's visitors looked Asian, with high cheekbones and eyes that were large and black.

Fortunately, Khoury hung on to his bizarre evidence. Eventually Bill Chalker was able to enlist the help of a group of friends that included scientists from the biochemical field. Chalker's circle of friends were interested in UFOs, but insisted their involvement would be on a strictly anonymous basis. Chalker became the public face of the UFO "invisible college" which called itself the Anomaly Physical Evidence Group (APEG).

The hair samples were examined by Dr. Horace Drew, who had worked for decades as a head research scientist at The Commonwealth Scientific and Industrial Research Organization (CSIRO), Australia's leading scientific research organization. As well, Dr. Drew co-authored the authoritative reference book *"Understanding DNA: The Molecule and How It Works."*

It was noted that the hair samples did not resemble any of the test samples taken from Khoury or his wife. In fact, according to Chalker's 1999 paper (*"Mitochondrial DNA Sequence Analysis of a Shed Hair from an Alien Abduction Case"*), the hair was "extremely thin and almost clear, and that further investigation...by high-resolution dark-field microscopy showed it to lie at the lower end of normal human hair thickness, and also to show a pronounced 'mosaic' structure, perhaps due to the near-absence of melanin."

Dr. Drew's PCR (Polymerase Chain Reaction) DNA profiling of the hair revealed that it came from someone who was biologically close to normal human genetics, but of an unusual racial type – a rare Chinese Mongoloid type – one of the rarest human lineages known, that lies further from the human mainstream than any other except for African pygmies and aboriginals. Even more bizarre, the mitochondrial DNA profiling revealed a rare Basque/Gaelic type DNA in the hair root, along with indications of the CCR5 gene deletion factor – indicating possible viral resistance against diseases such as HIV and smallpox.

The incredible finding of the DNA study seems to show human genetic manipulation on a scale not yet accomplished by modern science. There is also the disturbing indication that the "humans" that supplied the hair sample have been genetically altered to be resistant to sexually transmitted diseases such as HIV/AIDs. This is an interesting development considering the stories that have been circulated for decades about a possible human/extraterrestrial hybrid program being conducted by some unknown group(s).

HUMAN DNA

Even though Bill Chalker's book is titled *"Hair of the Alien,"* the DNA results, while extremely unusual, show that the owner of the hair is human. Caution must be taken not to jump to any conclusions on the origins of the hair that Peter Khoury found...no matter how bizarre the circumstances were prior to its discovery. That being said, there is no doubt that the DNA tests show that the owner of the hair is a highly unusual person whose genetic makeup is like nothing ever seen before.

In 2012, Chalker wrote in his Oz Files blog that the "Asian mongoloid sequence, found in the Khoury sample, is only found in the DNA signatures of an isolated group of people – the Lahu, who are limited to the region of the southern Chinese province of Yunnan, and the immediate regions bordering that locality – northern Thailand, Myanmar (Burma), and Laos." He also adds that ancient Lahu mythologies recount visits from the "sky beings" and today the region is rich with UFO and unusual light phenomena.

The question of human bioengineering, along with the centuries old tales of sexual interactions between humans and "nonhuman" entities, is maddeningly complex. With the exception of the occasional slipup, we are looking at something that has been very effectively kept hidden from us. Unfortunately, those slipups always bring up more questions than answers.

A TEACHER OF HUMAN SEXUALITY

Peter Khoury was no stranger to the abduction phenomena. The UFO-related incidents that he had vague memories of could have been just a small part of possibly many incidents that had been wiped from his memory. It is no wide stretch of the imagination to say that what happened to him in his home in Sydney did not just happen out of a vacuum.

Looking at Khoury's experience, it could be that at some time he had been tasked as being a "sexual guinea pig" or even an instructor of human sexual habits to young, non-human (or hybrid) entities, that are learning just what it is to be human. Given that shortly before his 1992 bedroom event, Khoury had been injured in a fight and was on medication, it is possible that his injuries, along with his prescribed drugs, had short-circuited any earlier mental programming, allowing him to not only be aware of his bizarre visitors, but to remember them afterwards.

Khoury recalled that when he violently resisted the "Nordic woman's" efforts to draw him to her breasts, she look surprised. As well, there seemed to be a telepathic communication to the "Asian woman" indicating that his reaction was wrong, that it wasn't supposed to happen that way. Was she surprised because Khoury had been previously "programmed" to respond correctly to the sexual stimulus, but instead responded in fear and violence?

As with other abduction scenarios, there were similarities with Khoury's experience. The women seemingly were able to instantly appear and disappear from his bedroom. Also, there was obviously some missing time with this particular event as Khoury was not able to recall how the two strands of hair ended up wrapped around his penis. This leads to another unanswered question...why were the strands of hair left behind in the first place?

A HIDDEN AGENDA

The UFO/Abduction phenomena have operated under the radar for possibly centuries, maybe even longer. Whatever the source of the phenomena, it has worked under the strictest of secrecy. As was mentioned earlier, there have been the occasional mistakes that have left suspicions of some sort of covert activity taking place. However, in regards to the hair samples left with Peter Khoury, they were not left behind by mistake. They were deliberately wrapped around his penis in such a way that they would cause pain and quickly draw attention. Considering the history of secrecy surrounding abductions and human/hybrid sexuality, this breach in protocol is indeed baffling.

The extremely bizarre nature and methods of the creatures involved with abductions suggests that traditional ideas about hybridization between humans and extraterrestrials are woefully inadequate. The events of July 23, 1992, show a shocking sophistication that goes far beyond the ideas and theories that have been entertained so far. One aspect of this sophistication comes from the discovery in the hair samples of the CCR5 deletion factor, which enables viral resistance to human sexually transmitted diseases such as HIV/AIDS, and possibly Smallpox. This mutation is relatively young, appearing possibly no more than 5,000 years ago.

The appearance in Khoury's hair samples may indicate that this particular mutation may be the result of bioengineering, and not natural evolution. This mutation could explain one reason that the "aliens" involved with human sexuality programs are able to interact physically with humans with impunity. This mutation may also be passed on to any descendants of these sexual interactions, thus producing hybrids that are more and more resilient to the potentially dangerous viruses that infect our species and ecosystem.

Throughout this article I have resisted the implication of "aliens/extraterrestrials" as the main player in the abduction mystery. I think that interplanetary visitors may be too simplistic to explain the mysteries surrounding UFOs and abductions. Considering that UFO occupants are often reported as humanoid in shape and behavior, it makes me think that unless the human shape is ubiquitous throughout the universe, the likelihood of an extraterrestrial race(s) looking almost exactly like us is extremely small. Yet, time and time again, we get sightings of humanoid UFO occupants. Why is this?

"Alien Hunter" Derrel Sims has suggested that we have never actually seen the extraterrestrial intelligence behind the UFO/abduction mystery. Instead, he says that the creatures that interact with eyewitnesses are biological constructions using earthly DNA so that they can safely operate on Earth. The so-called "Nordics," "Reptilians," "Insectoids," even the big-eyed greys, are bioengineered robots and not the true "Aliens."

The humanoids encountered have similarities that are often reported across the board. They are often described as having larger than normal heads and eyes. In fact the eyes are often said to wrap around the sides of the head. Their faces are seen as having high, prominent cheekbones and their chins are long and unnaturally pointed. Their skin color ranges from extremely light, almost white, to dark with an olive tinged color. Again, it is odd that beings with very large eyes, high cheekbones and pointed chins seem to predominate with these occurrences.

With these cases of high strangeness, there is no telling who or what is behind the UFO/abduction mystery. I cannot rule out time travelers or visitors from alternative realities because whoever they are, they have an intimate knowledge of humans that one would think would be beyond visitors from other planets.

The suggestion made by some abduction researchers that an alien race is trying to save their species by integrating human DNA into their own gene pool sounds like the plot for a science fiction movie. Experiencers who claim to have met human/extraterrestrial hybrids say that, for the most part, the hybrids look more human than alien. If an extraterrestrial species is trying to save their gene pool with borrowed human DNA, I doubt that they would be happy with the results looking more human than their own species.

We cannot dismiss, though, the disturbing fact that this program, whatever its motives actually involve, has possibly been in operation for thousands of years. This kind of long-term methodology will continue to stymie us until the time that the operators finally decide to reveal themselves. The question that we have to ask is, do we really want to know what is going on, or are we better off living our lives in ignorance to a possibly soul-shattering reality?

* Originally published in *"Screwed By the Aliens"* – 2018, Global Communications/Inner Light Publications

There are locations all across the planet alleged to be "window areas" where all manner of strange activities occur.

22.

THOMAS FERRARIO OFFERS MORE STRANGE ARTIFACTS TO BECKLEY AND SWARTZ

In a follow-up to the conversation between Ted Phillips, Timothy Green Beckley and Tim R. Swartz, we offer another interview with a UFO expert, Thomas Ferrario, who worked alongside Phillips in researching the Marley Woods region in Missouri for many years. The tales told by Ferrario include a horse spontaneously exploding and filling his stall with "horsemeat," as well as the mysterious appearance of large, 13-inch tracks on the ice covering of small ponds in the area — ice much too thin to support anything as heavy as the tracks would indicate. The following interview is from an episode of "Exploring the Bizarre," hosted by Beckley and Swartz, who begin with some background on Ferrario and his many years as a UFO researcher.

• • •

Swartz: Our guest tonight is Thomas Ferrario. Now Thomas has worked as a dive master, machinist and electrical engineer on projects in the United States, China and Bermuda. He's been an independent UFO researcher since 1969 to 1998, at which point, Walt Andrus, founder of MUFON, asked him to become a Section Director for MUFON. Later, he would go on to be Assistant State Director for Missouri MUFON. He then cofounded the MUFON dive team. He next joined Ted Phillips as his assistant in 2006 and later became part of Ted's SIU team. He assisted Ted Phillips on his Marley Woods projects and his Tetra Mountain Moon Shaft project. So, Thomas, thank you very much for being with us tonight on "Exploring the Bizarre."

Ferrario: It's good to be here. Right now it's good to be anywhere.

Thomas Ferrario

Beckley: I would agree. You know, I hardly know where to begin the conversation. I would assume that our listeners know a little bit about the background of Ted Phillips, having been an associate of Dr. J. Allen Hynek and involved in investigating over 400 physical evidence/trace cases where these objects had landed. And of course in recent years, he was staking out this rural area in Missouri known as Marley Woods where these mysterious phenomena and anomalies have been taking place for quite a few years, I understand. But why don't you start out, Tom, by giving us a little bit of background on yourself and how you ended up hooking up with Ted Phillips.

Ferrario: Actually, it started around 2000 when we had in Missouri an International MUFON Conference in St. Louis. I had always followed Ted's work and wanted to hook up with him. Speaking with him at the time, he found out I was an electrician so basically it was a case of I was the grunt. I had the opportunity to do some of the wiring and camera installation. It got to be that kind of relationship at first in Marley, which I was glad to do, you know. I might just throw in here, just a little announcement, the property owners at Marley Woods have an announcement to make, that they permanently want their ranches to be known as the Ted Phillips Marley Woods Research Center. So that's quite an honor. They want me to be part of that and help carry on Ted's work. I just feel it's quite an honor, it really is. And through the years Ted and I built a good relationship in Marley. And Ted decided that we needed more team members and so we formed a team and brought in several people. We brought in Adam Johnson, who is an outstanding videographer and now a producer in his own right, involved in productions on the SyFy Channel and other channels.

And we had Debbie Zigglemeier, who is the State Director for MUFON now, as a team member. She's also a Star Team member with MUFON. And I might just add that Rodney Dillard, from Branson, you might have heard of The Dillards, performers, he's also a team member. And of course we have one of Ted's great friends, who was an advisor to Ted for many years – we have Jacques Vallee.

Beckley: I noticed there's a photo of you and Jacques. I think it's posted on our site along with Ted. Now tell us just a little bit about what goes on – for those unfamiliar with the site – what goes on in Marley Woods? What have you perceived?

Ferrario: Well, I can give you a rundown here, a catalog. We don't have time enough to go into depth about anything. But basically in Marley Woods we have the light ball phenomenon; we have the unseen force phenomenon, strange animals. You just wouldn't imagine the gamut that that covers. Of late, we have ice tracks, we call it, and of course the light beam that we've all experienced out there on us. And then it gets into the real strange things, which we call the "horse barn incident." Which is really bizarre.

Beckley: What was the barn incident?

Ferrario: Basically what the horse barn incident was – one of the ranchers one day called Ted up and this actually occurred on a site, too, in Marley Woods, which is a northern site. Marley Woods encompasses two miles east and west and three miles north and south. So that gives you some idea of the scale that we run around on out there. Basically a rancher had his prize horse out in the barn. He was going to go out and do his chores and run the fences. He checked on his horse and his horse was fine, in the barn.

Thirty minutes later, he came back to the barn. The gate was knocked down. And this is one of Ted's strangest cases. The rancher called this in, and the only bad part about this whole deal was, he didn't want to talk to people or tell about this for six months. It was just so bizarre. But virtually what the rancher found when he went in the barn was what had been a perfectly healthy horse thirty minutes earlier – well, his horse now was turned into horse meat. It's on the upper loft area, it's on the gates, it's on the floor, it's just remnants of this horse. He said it looked like it had exploded in this barn.

So we went in and did some soil analysis, picked up samples. The rancher waiting until six months later was the problem, though. If he'd gotten us there earlier – we did see signs of blood and such in there. But this so disturbed this farmer. We were in there one time, and when we got back he had burned the barn down.

Beckley: Oh, my goodness. Now, what have you actually seen there? You mentioned about tracks in the ice?

Ferrario: Now that's a new phenomenon. To be honest, I haven't been out there. One of our new researchers, Rodney Dillard, and Ted, actually, had looked into this. It was one of Ted's last investigations. And virtually what you have, and this is high strangeness again, on small ponds you have tracks that approach 13 inches in diameter with more of a pronounced three-toe aperture on them. And the strange part of this is, you would think a 13-inch track, you're dealing with considerable weight there, you know. But this was on light ice; this would not withstand a person stepping on this ice. They appeared to be melted in the surface of the ice in parallel. They went across the ice, across the pond, and started coming up the bank and then that was it. They just disappeared.

Beckley: How many occasions? Was it just once?

Ferrario: Yes.

Beckley: Okay, I understand that you are involved in diving. Are there other anomalies going on in the ponds there?

Ferrario: There were quite a few. We got into cave exploration out there. You know it just covered the gamut of high strangeness. We could get into the animals. I was fortunate that I was with Ted when he had one of his – you know, when the researcher has experiences and becomes part of the research, that's really unique. And I was fortunate to be with Ted Phillips when he had his most astounding experience in Marley.

Beckley: And what was that?

Ferrario: That was the light beam phenomenon. What that really was was Ted and I were going down a long private road...

Beckley: When would this have been?

Ferrario: This probably would have been in around 2009. We were going down a private road on this ranch. Twelve hundred acres. A private road. It's gated in and out. We were coming out of there after a long night of surveillance. And all of sudden, this light beam from behind comes up and illuminates the whole interior of our truck to a degree of pure white light. I hate to do a parallel to "Close Encounters," but that's what it was. You could see the dashboard, the grain in the vinyl. I said to Ted at this point, "Ted, are you seeing this?" And Ted said, "Well, hell yes, I'm seeing this, Tom." So I slammed on the brakes and we jumped out. And this was the part Ted didn't like to talk too much about years later, because he always wanted to do the research before we made any statements. Well, we jumped out of the vehicle and we looked up at where the light was coming from. All we could see, for just a brief instant, was the remnant of – I could describe it as just like a blind closing or an aperture in a camera closing. And it was gone. That was something. That really set Ted back.

It really brought home that part of Dr. Hynek's opinion that Ted had a little bit of a problem with over the years. Because Ted was a nuts-and-bolts man. He wanted to get out and do the research. And this was really confirmed – Dr. Hynek, I might tell you, he had more or less thought that this was all "frequency-driven," this phenomenon. Basically, what that is, in laymen's terms, it means it's dimensional.

Beckley: Well, I know that, of course, for a long time Hynek was a nuts-and-bolts researcher as the majority of people were, including I guess myself, for a period. And then as things got stranger and stranger it became obvious that we had to be dealing with something that was more of a paranormal nature. And I know I talked to Ted about this a couple of times and he had spent all these years investigating the trace cases and looking for physical evidence. And then he slowly realized that the course of the phenomenon or the patterns involved were getting away from that and getting into other elements.

Ferrario: Yes, and that's the great thing about Ted's work. As hard-based as he was doing the true science, and, believe me, he beat that into us, about doing the scientific approach. But Ted was always prepared to go where the evidence, the science, led him. When this morphed into something more than structured craft, Ted went with the evidence. And that's why so much of the light ball activity just really blew him away. And you know Ted really hadn't seen anything himself until he got to Marley Woods.

Beckley: Now when did this first start developing? What year?

Ferrario: Marley Woods – Ted's been going there for twenty years now.

Beckley: And what was the last time that anyone visited there?

Ferrario: Probably, I would say, 2016. Maybe early 2017.

Beckley: Are there plans to continue the research there?

Ferrario: Oh, yeah. We're getting the team back together and we plan to continue Ted's work. We have the full assistance now of the ranchers. And, to be honest with you, I'm in contact with the property owners every other week. And they want us back down there. As soon as this damn virus gets burned out –

Beckley: Oh, yeah.

Ferrario: We're going to be down there.

Beckley: Have they reported anything happening since you've been there last? What's been going on?

Ferrario: This stuff seems to read your weakness. And when it knows we can't be down there, it basically goes crazy. One of the properly owners said that his grandsons were down there at one of the cabins on the southern side and experienced the amber lights just to a degree of intensity that they've never seen before. And even the strange animals we get out there have increased. So I just want to throw in what Ted and Jacques Vallee – they thought that Marley was so much more then even Skinwalker Ranch. Because you name it at Skinwalker and just when I think we haven't had that, something similar happens.

Beckley: Well, we did a book recently called "*UFOs Déjà Vu*," and it turns out that there are a number of these places. Of course there's the San Luis Valley in Colorado, where Chris O'Brien has been investigating all the activity there. There's the area around Sedona, Arizona, that Tom Dongo has staked out for many years there and has taken all these incredible photographs. Of course New England is quite a hotbed of activity on and off that Linda Zimmerman has been involved in and keeping track of. So there are these patterns. But Marley Woods certainly encompasses all of this.

Now, when this first began, was it more simplistic than it is now? I mean, did it start out with small lights and the lights continued to grow? I believe Ted told me that there was some phenomenon that was actually taking place inside one of the farmhouses?

Ferrario: Yes. The light ball activity has morphed and grown into just going into other branches. I could say the light ball activity goes back decades in this area. As far back as the ranchers can remember. They used to have picnics and go out at night and watch the lights –

Large orb photographed by Tom Dongo at the base of the Bradshaw Ranch.

Swartz: So you were telling us about the early phenomena there at Marley Woods, which consisted of amber-colored lights being seen.

Ferrario: Yeah, it actually started with the northern mystery light, which is a pretty normal phenomenon there. It's really regular, like every other night at least. It's in the same location. Goes vertical, it goes down, and extinguishes. That's usually the end of that sighting. But then we have the little white balls that go around and are the size of a softball. That was actually one of Ted's first sightings when he was out there. Ted was out there, a seasoned professional with a camcorder in his hand. And he was talking to the property owner, and Ted says, "Well, when does this stuff start?" And the property owner says, "Well, Ted, it started right there in the trees." So Ted stood there, watching this object. And the property owner told Ted, "Ted, I don't know too much about this stuff, but shouldn't your camera be pointed at the object?" Ted grabs his camera, and, seasoned professional that he was, of a three-minute sighting he got like thirty seconds at the end of it. Like Ted said, this stuff catches you and when you first see it yourself – from all the years investigating, when you start seeing all this – but yeah it's the light balls and then they progressed into larger ones. And like you mentioned, the amber lights are a much larger, intense source of amber-type radiation. Then you have not amber but yellow.

And the ones you really have to watch out for – you know you're dealing with a huge amount of capability and force – is the red lights. When you see the red light balls, you know you're in for an experience.

Swartz: Does anyone know how long this activity has been taking place in this area? The current landowners – is this something that has been in the family for generations? Or are they basically newcomers?

Ferrario: We've gone back close to one hundred years. If that tells you anything. Sometimes out there in a seven-acre field there's a little thing called "twinkle lights" where they get lawn chairs and go out there and watch it in the summer. There'd be thousands of little lights that cover this field. There's just incredible stuff like that, you know. And then the light ball activity. In the nearby town out there, there were two lawyers and there actually is a cemetery out there. There were two lawyers who thought this was all a joke. They'd heard some of the ranchers, they did some of their work. And they came out there one night just to poke fun at them a little bit. They're out in the cemetery about eight feet apart. All of a sudden – and I know that sounds like a bad joke – two lawyers in a cemetery at night. They start seeing eight of these light balls pop up from ground level out in the middle of the field, go into a formation, and one by one they instantly shoot about 200 feet between the two lawyers. They break formation and each one – like they're trying to show these guys something – and just go right between basically their heads out there in the cemetery. And no wind and no smell. Well, there wasn't any smell until after this happened, I suppose.

They can exhibit great force. One of Ted's primary cases that just shows the true nature of these things and again it involved a red light ball. There's a man who has a junkyard right down the road from the cemetery and he's got security cameras up in his car lot and has gotten a lot of footage over the years. But one night he and his wife were sitting there watching TV and all of a sudden the picture tube cracks. And what he described as coming out of this fissure in the crack in the picture tube, the glass became literally powder and shot out on the floor.

Well, they ran up to the television set to see this effect, and all of a sudden – they're in a two-level house, I might add, with a garage door in the bottom – all of a sudden that whole building just shakes. They get up to the window and look out. And he had had a full-size, four-wheel drive, Dodge pickup down there with the engine out, locked in gear, and the tires were flat. And this thing had been pushed up the driveway. He had a tow bar on the front of this truck. The tow bar went through the garage door and the grill of the truck was right up and pushing in the garage door. Now what has that kind of force? You know? And why?

We had five people who got out there and tried to push and budge this truck. And Ted said there just was no explanation for this.

Swartz: That's a lot of power there.

Ferrario: Oh, yeah. And as Ted said, is this something that they're trying to warn people? Or were they trying to beam this truck up or levitate this truck? And they misjudged the weight? There have been other cases where there's been cattle – typically the cattle are beamed up. There have been instances where these things have been dragged up and through the treetops. Just back and forth, back and forth, until they're just like a bloody pulp and fall down on the ground. Now, did they misjudge their capabilities or the true weight of these objects?

Swartz: That's interesting because that's a characteristic I've seen in other of these kind of like hotspot locations like Skinwalker Ranch – whatever this is, whatever these energies are – they seem to be extremely callous towards animals. There are a lot of mutilations. They seem to have it in for animals. However, it's almost the opposite when it comes to people. You very rarely hear of anybody ever being hurt to any serious extent at these locations.

Ferrario: Right. I'll be honest with you. We've had a little bit of that in Marley. There have been some side effects that haven't been that pleasant. But it's always I believe an artifact of their craft or propulsion system and I do not believe it's intentional. I really don't, as you stated.

Swartz: When you say "craft" or "propulsion system," what do you mean? What do you think is going on? Do you think that you are dealing with some kind of extraterrestrial intelligence that's using physical craft? Or what?

Ferrario: We really believe that it's actually both. The determination has come out of this over the years, that Ted thought he'd never say, that him and Hynek – Hynek was trying to impress this on him and finally Ted came around, but what's going on out at Marley – I've put my stock in J. Allen Hynek and Ted, over anybody else, until they prove me wrong. But this is all interdimensional and not interplanetary. And that's the one thing, I can tell you, that they came out of this with. There are physical, structured craft that come through. We've had black cylindrical craft out here. Property owners were going home one night and saw something that approaches over a football field in length, a large, cylindrical, jet black craft. And at this point they've actually seen the light balls coming out of one end of this craft.

And that was the first and only time we've ever had a structured craft sighting in Marley. It's all together out there. It's all related.

Swartz: Well, talking about Ted himself. What kind of man was he? Do you have any idea what got him interested in this subject to begin with?

Ferrario: I don't know how many people know this, but Ted had an instinctual interest in this subject when he was a child anyway, but his father introduced him at the time – he was born in Sedalia, Missouri, in 1942. And his father had heard a report on the news and knew that Ted would be interested.

What it was, it was a WWII fighter pilot that was reported on the news that night that he had had a close encounter, a UFO sighting. And his father went and got Ted and took him and Ted interviewed this man. And the fighter pilot told him about seeing an amber-colored, 30-foot disc a hundred feet from the cockpit of this fighter plane during WWII. And like Ted said, it doesn't get any better than that as far as an eyewitness. So that's what got Ted keyed into this research. That's what got him started, along with other things, with the physical trace aspect, the nuts-and-bolts.

Beckley: Well, Socorro. When he was about 20 years old he took a trip down to Socorro and that's where he met Hynek.

Ferrario: Right. He met Hynek and another gentleman named Ray Stanford was there. You may have heard of him.

He's a good friend of mine. He wrote *"The Saucer in the Pentagon Closet"* I believe. But the three of them actually did Socorro. It's so sad. Before Ted passed away, Ray and Ted were going to write the *"Socorro: Closed Case"* book on that. It had never really been done sufficiently, Ted felt. It's a shame that will never be done now.

ALIEN ARTIFACTS

Arrow Shaped Object Uncovered in Ukraine

During the summer of 1947, construction workers sifting through the ruins of Kiev, Ukraine, stumbled upon an extraordinary object at the location where the Kiev conservatory was located.

The object was buried 5-6 meters underground and was described as a silvery, seamless arrow-shaped object, about 3-3.5 meters in diameter and 5-6 meters long. Its backend was evenly cut, apparently torn from an even larger object. It had narrow attachments on its sides that gave the object the appearance of an arrow.

The finding was excavated and military field engineers were called out. The site was immediately cordoned off. It was speculated that the object could be some foreign missile. Secretly, the object was loaded upon a truck platform, covered by a tarpaulin and transported from the Ukraine to a Russian location named Podlipki, a place located northeast of Moscow.

(later this city was called Kaliningrad and now it is the town of Korolyev in the vicinity of Moscow). Once there, the first Soviet missile research center was established.

The strange object was rigorously studied and it was established that it had been buried for several thousand years (approximately 3,000 to 5,000 years).

It was also discovered that the object had been originally found in the 19th or early 20th century by the famous archeologist Vikentiy Khvoyka.

The governor of Kiev was invited by the Tsarist Police and authorities to examine the finding. He attentively inspected the object, but ordered buried again since he did not dare take responsibility for something that was impossible to explain or study during that time.

In 1947, Soviet scientists and pioneer missile designers were fortunate to be able to study the re-discovered object with all the equipment that was available to them in their laboratory in Podlipki.

Upon examination it was established that only the front cockpit section of the craft had been found. Originally the craft must have been around 11-12 meters long. Unfortunately the engine compartment was never found. The front section appeared to have been had been cut away as if by a high intensity laser.

The craft was obviously not of terrestrial origin and of non-human manufacture, since it had two small chairs for two very small pilots that had been approximately 1.2 meters in height (the cabin was empty and the bodies were never found). Sophisticated equipment, control instruments with levers, buttons, etc and panels were found inside.

The strangest thing was the unknown language found inscribed inside the cabin. After deciphering the language it was determined to have been ancient Sanskrit.

Sergey Pavlovich Korolyev, who later became an academician and a famous Soviet missile designer, headed the group that was created to study the discovered alien spacecraft.

Reportedly the craft was extremely advanced above the level of science and technology of the USSR in 1947, and very little of its structure could be understood and almost nothing could be replicated. However, extensive reverse engineering of the alien craft was accomplished in 50's, 60's and 70's.

Eventually, the back-engineering helped in the development of Soviet missile and space technology, including the design of metal alloys, instrumental design, control systems and some construction elements.

10 years after the objects discovery in Kiev, Korolyev helped launch the first satellite (Sputnik) to space...and in 1961 –Yuri Gagarin was the first human astronaut.

To the amazement of the West, this was done by a country considerably behind the United States in aviation, space and missile technology.

This story was leaked by Mr. V. Sukhoveev, resident of Kiev whose father (a known linguist) was called to decipher the inscriptions found onboard the object.

Source: Anton Anfalov, Local Press, V. Suhoveeva's testimony, Lenura Azizova

23.

HERCULES AND THE ALIEN ARTIFACT
By Hercules Invictus

I: IN SEARCH OF: An Alien Artifact

My objective was simple: I wished to obtain a celestial artifact that would allow me to collect positive, powerful and transformative energies, pool them, concentrate them and then apply them toward countering a negative, disruptive and hateful astral current that increasingly threatens our world.

Through the vehicle of meditation my mind ascended toward a golden structure suspended in the heavens above New York City.

To some, this vast complex manifests as a galactic vessel of extraterrestrial origin that is staffed by evolved alien visitors from other worlds. It is believed to exist on many levels of reality simultaneously, but it can be fully experienced, in all its magnificence, only by Etherians on their native Etheric Plane.

To others it appears to be an Astral City, one of several stationed in our skies, that monitors the status and activities of certain embodied souls, both Wanderers and Walk-Ins, who volunteered to complete important tasks whilst incarnate on the earth plane. The city provides instruction, guidance and support to its volunteers whenever necessary and/or possible. It also acts as a healing center and way station between the death of the earth-bodies of their charges and the choice of a suitable vehicle for their next incarnational assignment.

To me it serves as both of these things, and much, much more. It has assumed various semblances over the years, yet it remains the same in purpose and in spirit.

Hercules Invictus

II: ONCE UPON A TIME: In the Realm of Dream

When I was much younger, I would sometimes find myself on an elevated train much like the ones we had in New York City. I would always take it to the last stop. The area outside the station was much like the residential areas in the outer boroughs of my youth, but here it was always spring, daytime and very sunny.

Usually I would realize that I was dreaming at this point and would decide upon my destination. There was a very steep hill across the street from the exit, about half a block to the right. At the apex of this hill, to the left, was an interesting city, and a peculiar institution about halfway to the top, but to the right. Reaching either involved some strenuous climbing.

Onwards and upwards!

At some point I would reach an ornate metal fence, and ultimately an imposing gate that, if unlocked and open, allowed me to enter the grounds that surrounded the Hall of Akashic Records. There was a well manicured lawn, plus lots of expertly sculpted trees and bushes as far as the eye could see.

The Hall itself reminded me very much of a university building, and a well-stocked library once within its massive doors. The Librarian was an ancient and wizened man with a long, and often unruly, silver-white mane of hair and beard.

If I presented him with a request for insight or information he would scamper off and seek it, quickly returning with his arms full of books containing lots of useful material.

The contents of these resources could be explored in the Reading Room or experienced through other dreams. There were pros and cons to each of these approaches. In either case, the results of the inquiry were not often accessible to the mind once awake. Information would be released to my brain, if needed, through flashes of insight and/or strange synchronicities.

The city atop the hill was similar to the waking world in many ways and it always felt like I was coming home whenever I chose to visit. There were cycles of day and night plus the passage of seasons. I had friends here and family as well. They all aged, and changed, with the passage of time, as did the city. The one great difference between this place and the world I found myself in when awake, was that there were no discernable sexual, racial or cultural tensions between the diverse people that lived here. There was instead total equality and total respect in their interactions and institutions. Difference and diversity were embraced and celebrated.

During the passage of years I experienced good times and bad times when visiting this place. People lived and died. There were shifts in population and disturbing fluctuations in the economy. But the shared values of the residents remained strong and guided their efforts. They believed that they were all in it together and together they could make it through, whatever the nature of the challenge they faced.

The city was a model of how things could be in our world as well, if we chose to embrace the siblinghood of humanity and put our faith in ourselves and in each other.

III: BACK TO THE PRESENT: The Akashic Records

What was once a physical and very traditional edifice for higher education on landscaped grounds now appears as a vast and futuristic living computer. The ancient Librarian went through several iterations over the years before manifesting as an AI with holographic capabilities. In my dreams, or through my meditations, I sometimes re-experience old forms, but my experiences are always expanding and evolving the options and possibilities available to me when I am here.

I no longer need words to express what I need from it. The hi-tech, sci-fi device immediately knows, and quickly responds. I see several images: shapes, colors and symbols, first apart, then in sequence, then superimposed. And I get it, receiving and fully understanding the message on the spot. No more dream distortions or delays!

Now it was time to proceed to the city atop the hill, now floating in the clouds! It too had evolved in my perception over the years.

IV: THE CITY IN THE SKY: My Return to Atlantis

In neo-Theosophical lore it is recounted how the Goddess Liberty herself erected a vast Temple complex honoring the Sun and the Twelve Solar Hierarchies on and around what is now the Greater New York City Area in the halcyon days of Atlantis, . The Central Altar is in New York Harbor, where the Statue of Liberty now stands. Though there were several such complexes on our globe throughout our history, the one past the Western Gate of Atlantis was our planet's primary focal point for Gaia.

ALIEN ARTIFACTS

The physical structures are said to have been destroyed during the Cataclysm but the Temple itself still exists and has remained active, hovering over modern Manhattan on the Etheric Plane, channeling the powerful solar energies that guide our planet's evolution.

Since the days of myth, ancient mariners in Europe have braved the unknown and sought legendary treasures across the Atlantic, including the Golden Fleece and the Golden Apples. Both have strong Solar associations and are linguistically interchangeable. Some have suggested that these items represented actual gold, copper, tin or even psychotropic plants. And indeed, there is much evidence to support each of these theories. The ancients generously left behind ample, but enigmatic, evidence of their visits.

In Solar symbolism the voyage westward, in the direction of the sunset, releases you from the habitual patterns of your old life and promises a fresh start, a new beginning, a rebirth.

Those immigrating to the United States seek the American Dream: life in a paradise where liberty reigns, the streets are paved with gold and there is equal opportunity for all. It is no accident that the Goddess Libertas looms large near their once-point of entry, wearing a solar crown, holding aloft a torch to keep the dark at bay and grasping a Book of Law. The Statue of Liberty is a symbol of freedom and was gifted to us by France, where the Flame of Liberty was secreted after Atlantis was swallowed by the waves. Life, liberty and the pursuit of happiness are declared unalienable rights in our Declaration of Independence, and the American government was created to protect them.

**From the Hall of Akashic Records:
A symbol of Cosmic Consciousness,
protection from all ills and the power
of personal responsibility.**

Some come here seeking freedom from religious persecution. The melting pot of people and cultures exchanging information, coupled with the freedom to think and communicate, have given birth to many new forms of spiritual expression. Theosophy, the I AM Activity, Celestial UFO Spirituality, Spiritualism, New Thought and the New Age Movement all emerged from the cultural diversity that defines us.

The Temple complex is an expression of the city atop the hill of my childhood dreams, or vice-versa if you prefer. No doubt it will continue to evolve as my own understanding expands over time. In the present I am guided to the Temple of the Fourth Solar Hierarchy, and am handed a gemmed staff by the attendant priests, who seemed to be awaiting my arrival. This particular Temple is dedicated to the constellation Taurus and to the Creative and Generative Powers. It is themed with Minoan, Mycenean and Atlantean motifs.

I recognized the item as the Scepter of the Shapers. Lore on the Crown and Shield of the Shapers (aka Elohim) can be found in twentieth century occult literature, if you'd care to look. Directions for their use are provided in the books but mastery comes only with practice. Noting the placement and color of the gems, I deduced how to best employ this particular alien artifact.

Leaving the city, I made my way toward what appeared to be a roiling patch of disturbance in the sky, fast approaching.

V: TYPHON: A Primordial Being

There have always been forces opposing its ideals but the city atop the hill, in all its diverse guises, was being threatened as never before by a shadowy, predatory parasite. In my perception it appeared to be like storm-clouds menacingly assembled. It looked like a chaotic amalgam of beasts, including a gigantic hydra with countless horrid heads. It dominated my field of vision and generated an aura of tremendous terror.

It whispered hate, hissed suspicion and spewed distrust. It enveloped me in its inky vastness and I felt isolated but not alone.

I began to feel victimized and grew extremely paranoid. I was convinced that I lived in a dangerous world where all manner of evil enemies were out to get me. I wanted to lash out blindly and I wanted to hide somewhere where I could never be found. The many other people I perceived, all around me but somehow out of sight, perhaps we could connect or team up. But who could I really trust? Perhaps they were all hypocrites and sly deceivers who meant me and mine much harm.

The Beast promised protection, safety in numbers and the delivery of brutal judgments and swift retributions against our many foes. I was filled with righteous fury. It then offered hope for a brighter future and a better world once all our enemies were humbled, humiliated and destroyed. The entity then allowed me to experience what that would feel like. I saw its vast army and felt elated, confident and strong.

And I recognized the Beast. This allowed me to dispel his seductive reasoning.

This was Typhon, the Father of Monsters in Greek Mythology. In the old tales, as a child and champion of our Earth, he single handedly drove the Olympians away from their Divine Abodes in the Heavens and into Egypt and the Near East for a space, where they assumed diverse cultural disguises until they could defeat him.

Hitting Typhon with my staff seemed rather pointless and it would place me firmly in his grip, an instrument of his will. Yet this is what the clerics of the Bull from Heaven provided me with. Declaring him my enemy and attacking his forces, both human and non, also locked me into his 'us versus them' mentality and, if I indulged in this, I would accomplish naught more than perpetuating what seemed to be a state of unceasing conflict.

If the Father of Monsters was quickly devolving us into a condition of increased savagery and tribalism, then the fate of our entire civilization was also at stake.

And if Typhon was acting as an agent of the Earth herself, then there were serious environmental issues that needed to be immediately addressed by humanity as a whole. Perhaps Typhon was preparing us for our coming round of human experience. Perhaps in his estimation, he was the Harbinger of Sudden and Much Needed Change.

There was still much to contemplate and comprehend. It was definitely time to take action. I knew that I must devote my remaining time and energy to the areas of greatest concern.

VI: THE SCEPTER OF THE SHAPERS: A High Tech Club of Hercules

I recalled that the weapon of Hercules, though often described as a Ropallon (Club), was also sometimes described as a Kladhos (living branch). There was even an ancient account of it being planted and springing to life as an olive tree.

Recalling the symbols I received from the Hall of Akashic Records, I called them forth and configured them around my floating form: an Eye, rayed like the sun; the

Heavens at night, as conceived by humans; a Thunderbolt, symbol of Divine Authority; the colors gold, white, red, blue and purple. When I returned to waking I would design a symbol to remind me of my commitment.

No, Typhon would not threaten the people atop the hill, floating in the sky or in their heavenly abodes.

I held the staff aloft.

The blue gem on the scepter lit up and the thick tendrils of Typhon started to withdraw. I uttered the word Vision.

The golden yellow gem on the scepter lit up and the darkness of Typhon in my immediate area grew dimmer. I uttered the word Gnosis.

The rose gem on the scepter lit up and the Typhon's air of oppression substantially lessened. I uttered the word Love.

The clear gem on the scepter lit up and I was personally clear of Typhon's malign influence. I uttered the word Creativity.

The green gem on the scepter lit up and the presence of Typhon seemed to withdraw even further away from me. I uttered the word Challenges.

The golden orange gem on the scepter lit up and the influence of Typhon seemed much less substantial. I uttered the word Leadership.

The purple gem on the scepter lit up and Typhon seemed no longer a threat. I uttered the word Structure. Rested, refreshed and at peace, I returned to the waking world.

VII: A CALL TO ADVENTURE: Argonauts Assemble

Another day another Astral journey. From Mount Olympus I blow into the Horn of Summoning, another celestial artifact from the days of Atlantis, and watch the skies to see who will heed the cry and respond to the call.

Perhaps it will be you dear reader, and you can then join us on our latest Argonaut adventure.

© Hercules Invictus

• • •

Hercules Invictus is a Lemnian Greek, a proud descendant of Argonauts and Amazons. He is openly Olympian in his spirituality and worldview, dedicated to living the Mythic Life and has been exploring the fringes of our reality throughout his entire earthly sojourn. For over four decades he has been sharing his Olympian Odyssey with others.

Having relocated the heart of his Temenos to Northeastern New Jersey and the Greater New York Metropolitan Area, he has been establishing his unique niche locally and contributing to his community's overall quality of life in any way he can. Hercules also recruits Argonauts to help him usher in a new Age of Heroes.

Hercules currently hosts several podcasts, has written numerous articles, has published two e-books on Kindle and has contributed to about twenty paranormal anthologies. He frequently conducts Olympian Workshops and serves as a Guest Speaker on multiple platforms.

For more information please friend him on Facebook https://www.facebook.com/hercules.invictus.7

Or check out his Podcasts:

www.blogtalkradio.com/herculesinvictus

Articles:

https://themagichappensnow.com/author/hercules-invictus/

Author Pages:

www.amazon.com/Hercules-Invictus/e/B07L8J3PQ7

www.goodreads.com/author/show/16042735.Hercules_Invictus

Objects that appear to be natural or manmade are often seen to fall from the sky. But what if some of these objects were deliberately sent by an unknown intelligence?

24.

FREE FALLING
By Tom Hackney

Alien hunters come in all shapes and sizes. Some wear long-flowered dresses, weigh two-hundred-and fifty pounds and worship the deceased gods of Egypt. Others wear lab coats, have advanced science degrees and look for extraterrestrial radio-waves. The latter can sometimes be heard saying "Science!" in their sleep. None of them look like movie stars. Me, I'm what you might call an *existential detective*, someone who looks for advanced intelligence in the world of probabilities, where reality blurs and bows down to intelligences too enormous and complicated to contemplate.

I've learned a few things about these powers from observing some of their handiwork. I've watched twenty-one comets impact the back side of Jupiter, which did in fact happen over six days from July 16 to 21, 1994. They called the bogey Shoemaker-Levy 9, as if it were just a single comet that exploded with the force of fifty million hydrogen bombs. Who knew – who wanted to know? –that SL9 was actually twenty-one separate comets, designated "A" through "W" (I and O were not used), especially since the 21st century just happened to be right around the corner. To the average astronomer at the time, nothing could have been more irrelevant. That quaint notion about comets auguring things – usually quite unpleasant things – is just silly superstition from bygone eras. It's certainly not science. For the existential detective, however, such coincidences are the very things that keep the bread buttered.

I've watched a five-inch by eleven-inch meteor pulverize a car taillight five inches by twenty-two-inches without significantly damaging its chrome borders. It didn't seem to occur to anyone as strange that this one-in-a-billion event occurred three days *before* NASA-Ames commenced the world's first publicly-funded "Targeted Search," or SETI project, to search for extraterrestrials on October 12, 1992. Falling rocks from the sky augur major historical events, too. It's a human thing. The High Resolution Microwave Survey was perhaps a million times more powerful than any SETI project before it. More radio-waves were collected and analyzed as being of intelligent creation in the first minute than had been accomplished in all the previous fifty privately-funded SETI projects combined. October 12, 1992 was also the 500th anniversary of Christopher Columbus's discovery of America, the "New World." Nice symbolism. No need to be burdened with the horrific history, though. It's not science. It wouldn't occur to anyone with a PhD after his or her name that NASA's alien hunt might have seemed impolite or threatening to any of the new worlds being searched for.

But what is reason in the face of dreams? Hang in there, gentle reader, I think we can sort all this out in a way that will please … well, nobody, really, but it'll be a solution we in the alien-hunting business can all live with. *Live* being a relative term. You see, the license plate of the car struck by meteor, 4GF-933, provided the month and year Shoemaker-Levy 9 would appear, which was March 1993, or a little more than five months later. Then there was the highly curious, if hard, cold fact that the car's owner just happened to turn eighteen on October 12, 1992. "All grown up are we?" These are the little things that keep me going.

A car in Peekskill, NY, that was struck by a meteorite on October 9, 1992.
(Stuart Bayer - *The Journal News*)

Just when I thought things couldn't get any stranger, they did. Turned out the fireball from which the Peekskill meteor was honed to its perfectly fitting size *began* its atmospheric flight very much adjacent the National Radio Astronomy Observatory in Greenbank, West Virginia. The NRAO just happened to be a major player in the High Resolution Microwave Survey. On its northeasterly journey up the east coast, the lime-green (what else?) fireball passed Washington D.C., which just happened to be the midpoint of the fireball's 700 km (1.) known flight path. Wasn't it odd, though? This was where the fireball began to break apart into some seventy fragments, according to photographs and films made at a little before 8:00PM. It was one of these fragments that terminated through a car's right taillight parked in Peekskill, New York, a small town about twenty-seven miles north of New York City. The happy recipient of the meteorite, the girl turning eighteen on October 12, later sold her 27.3 pound meteorite and damaged 1980 Chevrolet Malibu to a collector for a total of $75,000. Nice birthday present.

The more I looked into this meteor event, the more ironic and the chattier it became. Take the word "Peekskill", for example (please). Nothing like a small demonstration, or "peek" at someone's *skill*, eh? Of all the towns and gin joints in America ... you know the drill. The thing. Then there was the toy ape hanging from the car's rearview mirror. Honestly, who could have resisted that one? And the rather distinctive fact that it was the *right* taillight, not the left, that had been tagged within a millimeter of its life. NASA-Ames's HRMS hypothesis – "ET exists" – just dangled there. So did the reply: "*Right*, Ames!" Drums from the deep. If I'd had an aboriginal pointing bone handy I'd have chanted, flapped my arms, jumped up and down and used it, but who'd have listened? Being an existential detective can be a real bitch sometimes.

Does anyone really think the entity or entities that caused this ride around in flying saucers? No way! Those contraptions known as UFOs or UAPs just don't cut it, I'm afraid, zipping here and there as they please, but crashing all too often. Good enough to get from here to the moon, maybe, but beyond that I wouldn't get in one of those things even if the money was right

And yet someone is calling the shots here, making the rules of non-engagement that UFOs and their occupants follow. Otherwise, we'd probably be their slaves. Maybe we'd have been wiped out ages ago. Make no mistake, the Earth is a major prize, glittering in space like the priceless gem that it is. Someone is protecting us, keeping the mawkish human parodies at bay. My money is on those I call The Authorities, or The Landlords, as Arthur C. Clark called them in *2010: A Space Odyssey,*" the ones who caused the planet Jupiter to become a second sun, as a lesson to us (in the novel, you'll recall, the human race had been about to launch into World War III).

With Vladamir Putin now threatening to use nuclear weapons if the free world doesn't get off his back and let him rape and pillage Ukraine for its land and resources, maybe we're already there. Is China's takeover of Taiwan, that rich, juicy morsel, next? Seems it's back to the bad old days, when conquering and subjugating people was what powerful rulers did to get their money's worth, and because they could. Things seem to be going just as Shoemaker-Levy 9 might have predicted, in the "21st Century."

ALIEN ARTIFACTS

We're supposed to be better than this.

Of course, I can't *prove* anything. Everything is statistics and uncertainty, as scientists know very well, when it suits them.

1. "*Nature*" magazine July, 1994

* Tom Hackney is the author of "*The ETI Grail*" (2012) and "*Alien Memos*" (2018), about meaningful interfaces and communications from an advanced non-human intelligence, beginning in 1992 three days before NASA commenced the High Resolution Microwave Survey (SETI) project. Tom's articles have appeared in Fortean Times, MUFON Journal, 34th Parallel, Hack Writers, Adelaide and Fate magazines, among others.

25.

THE FAR SIDE OF THE LOOKING GLASS
By Scott Corrales

This case takes us from subequatorial Brazil to the Caribbean, where in 1934, a teenager identified only as "Julio" became the protagonist of an episode that would scar him for life. It was first investigated by Puerto Rican ufologist Sebastián Robiou (mid-70s), and then re-investigated by Salvador Freixedo (late '80s). The witness has since been interviewed once more by Magdalena del Amo-Freixedo (1997).

GO FLY A KITE!

One morning, while flying a brand new kite on a slope outside the city of Mayaguez, Puerto Rico, before going off to school, young Julio was startled to see his kite being sucked in by what appeared to be an air-pocket or vacuum of some kind. He pulled on the string and noticed an inordinate amount of resistance from the wayward kite. Upon looking up, he was amazed to see a ball, "like a ball bearing," but measuring some twenty feet across and having the same coppery hue of a BB. A light issued from the object and he felt himself being raised into the air. Before he knew it, he was inside the strange flying object.

"On one side, I saw a girl," Julio indicated during the interview conducted by Robiou, "and on the other was a guy looking at some sort of giant emerald. He wore a tight-fitting olive-drab suit that looked like plastic. I couldn't see his face because he was minding the device. He gestured at the girl...the girl had a pinkish complexion and wore a silvery suit. She was small, like one of our six year-old girls, with platinum blonde hair. I don't remember the color of her eyes."

Julio explained that the child was holding his kite in her hands, and that he made all possible efforts to tell her that it belonged to him. The girl not only did not surrender the kite, but instead gave him a small box, from which images could be made to appear. He did not remember how, but the object returned him to the place from which he had been collected, and returned him abruptly to the ground. He suffered a sprained ankle as a result of the experience—but he had the curious little box with him.

THE STRANGE BOX AND ITS 'CONTENTS'

Further details would emerge during Freixedo's re-opening of the case. "Julio," now a hardened man in his early sixties, informed the Spanish ufologist of the ultimate fate of the little box he'd been entrusted with.

The box measured some 20 x 20 x 20 centimeters, and when its "user" placed his or her hands upon it, a "kind of vapor made up of lights" would spin on its surface, causing an entity—as small ape-like creature no more than one meter tall—to appear in the room. According to Julio, the entities materialized in such a manner would not speak and appeared to be surprised to find themselves in an alien environment.

The girl-child on the strange object had successfully caused the "little apes" to return to their native surroundings or "back into the box," as Julio put it. Only the hapless boy was not so good at this final aspect: the diminutive simians would materialize and vanish at breathtaking speed out the window, many times in the presence of his classmates who had asked him to perform the "neat trick" with the box. The aported entities were not at all pleased, claimed Julio, with their new condition. They would frighten children and dogs, and appeared to prowl the surroundings of Julio's family's house.

"Believe me," he told the researcher. "I would just like to die. I'm tired of seeing strange things."

The supernatural primates had apparently been the source of a number of mysterious deaths which had occurred in his corner of southwestern Puerto Rico over the decades.

When Magdalena del Amo-Freixedo reopened the case as part of her book *"Abducciones"* (Bell Book, 1998), a further wrinkle appeared which has a direct bearing on this article.

LIFE'S A BEACH

Now willing to go on the record by his real name—Juan Rivera Feliberti—he explained to Del Amo-Freixedo that his contact with the alien "girl" had not stopped after the incident of the wayward kite. Many years later, now a married man with children, the experiencer moved from Mayaguez to Sabana Grande, P.R., and took his family to the beach one day. While the children frolicked in the water, "Julio" decided to go fishing. He suddenly realized he was not alone: a beautiful woman had appeared right in front of him. A wave of remembrances washed over him as he realized her blonde hair was identical to that of the girl in the odd circular vehicle so many years ago. He asked her where she came from, and she allegedly replied "from far away, from the stars..."

Male figures soon appeared, clad identically to the one he remembered seeing back in 1934. "They were the lady's companions...they were identical to the one I'd seen as a boy. Suddenly, I don't know what she did, but she was completely naked. She didn't tell me anything, but I understood in my mind that she wanted to have relations with me. I didn't want to...I wanted to run away. Besides, my wife could catch me if she happened to come around."

Although hesitant to describe his unusual experience to a female investigator, Del Amo-Freixedo eventually convinced him to elaborate. Uncomfortably, "Julio," now in his seventies, continued the story: "Look, I didn't want to at first, but you know how it is. I was young and the woman was very good-looking. She began caressing me all over, and we ended up like men and women do when they're both unclothed."

"Julio" bashfully added that his alien lover's body was not exactly like that of a human female: her breasts appeared to be placed lower on the torso and her pubic area was hairless. He made the curious observation that her skin, while soft, was somewhat scaly. These anatomical differences did not deter him, however: "We [had sex] several times. I think four. Back then one was full of energy and recovered quicker." In subsequent years, he would return to the scene of the events in hope of seeing his unusual sex partner again, but never did.

As if to bring the events in the long, strange life of Juan Rivera Feliberti to a full circle, at around 3:00AM one day in 1995, he saw the same girl who'd stolen his kite once more, standing outside his house.

WHERE ARE THE USUAL SUSPECTS?

Regardless of whatever stance we may have regarding the UFO phenomenon, and provided that we are willing to suspend disbelief, the information which can be gleaned from these cases is of considerable interest: absent from the scene are the Greys, Reptoids and Nordics that seem to populate the abductee chronicles. Instead we have beings of an entirely different taxonomy engaged in an operation or mission that appears to be taking place largely within the confines of Brazil, the South American giant. The commonalities of the experiences—the oily liquid applied on the abductees, which serves as antiseptic and aphrodisiac at once; the beverage that relieves human discomfort; the choice of intercourse rather than artificial insemination—link them together while separating them from the coldly clinical abduction phenomenon in the northern hemisphere.

The fact that this libidinous aspect of the UFO phenomenon appears to have a strong preference for Brazil has led to jocose comments on the appeal of Brazilian virility to non-human intelligences. The fact remains that somewhat similar situations have occurred elsewhere in the world and in our own country as well.

In October 1974, oil worker Carl Higdon took a day off from work and went hunting near Rawlins, Wyoming. Coming across an elk (an astonishing piece of luck in itself on the first day of hunting), Higdon pulled the trigger on his rifle only to see the bullet issue from the weapon in slow motion and land fifty feet away from him. To his astonishment, the hunter realized that time was standing still all around him and that a chinless, jawless alien being was looking at him. Higdon was apparently abducted and hooked up to strange devices aboard "a cube-shaped UFO." The hunter believed the reason for his return to Earth by his captors was that he had had a vasectomy performed a few years before the abduction and was therefore useless for the "breeding program" that his captors appeared to be pursuing.

Scott Corrales leads the way in the investigation of strange UFO encounters in Hispanic nations.

26.

COMMAND SERGEANT MAJOR JAMES NORTON: I HAVE PROOF OF AN ALIEN UFO!

By Paul Dale Roberts

Way back in September 1977, during the Joint Attack Weapons Systems Test (JAWS) at Fort Benning Georgia, the entire base witnessed what appeared to be a UFO invasion. As much as 1300 troops were involved in this incredible event. Most were left with severe psychological trauma and "missing time" gaps.

Command Sergeant Major James Norton, who is now stationed at Fort McClellan, Alabama, was a buck sergeant (E-5) at the time when he witnessed an eerie encounter with UFOs and the military on September 14, 1977. He was sent to the range just before midnight along with around 1200 to 1300 troops to perform live fire with their new weapons systems called JAWS (Joint Attack Weapons System). The Secretary of the Army at the time was actually at the range too!

Command Sergeant Major James Norton remembers clearly when they started the live fire training. All of a sudden there were large orbs way up high in the night sky above them. The next thing he witnessed was what appeared to be lasers crisscrossing through the night sky. The command of "cease fire" was heard shouted out towards the troops, because there was something down range. Military helicopters appeared suddenly out of nowhere. Explosions were heard, and complete madness had broken out around him! Orbs in the night sky were changing a variety of different colors, a lot like the kind of orbs that are being seen around the world today! Then a triangle-shaped UFO showed up. The soldiers around Norton were in complete shock and awe at this spectacle that was now suddenly in front of them.

After this incredible incident many soldiers became very ill. Every soldier who witnessed this crazy event was briefed that if they were to EVER disclose any information in reference to the night of these bizarre events, they would be court martialed immediately and sent to the federal prison in Leavenworth. When Norton eventually obtained his medical records it disclosed that his illness was of unknown origin, his temperature was 104, and he was placed in a steel tub of iced water for two days, to help bring his temperature down.

Norton had a memory lapse or, as we know it in "UFOs speak," is also called "missing time". He could not remember how he was on the range and then later found himself alone in the woods. There was an awful lot of time disorientation. During a hypnosis session, he learned that he was actually abducted and taken aboard an unknown craft. Medical experiments were conducted and samples were taken from his body.

After the incident two jets came in that appeared to be white washed; no numbers or identification were on the jets. The jets were there purely to collect UFO crash site debris!

UFO PROOF

Norton picked up a metallic piece from the crash site that had/has strange hieroglyphics on it. This metallic piece cannot be burned. A hacksaw will not cut it either and, if you bend it, it will just go back to its original shape. Norton has this metallic piece apparently BURIED AT HIS PROPERTY!?

Norton is now willing and ready to show this evidence to some credible UFOlogists and to the media! Norton claims he has 100% proof of UFOs being real with this solid piece of evidence that he has held onto since the incident itself.

MILITARY HELICOPTER SHOT DOWN

The military helicopters that were shot down during the incident were never   reported to the FAA. He was also burned from a mysterious red substance from this incident!?

Norton is now stationed at Fort McClellan, Alabama, and describes how they experiment with biological and chemical substances on this base and that something else is going on there. It seems that our military is test flying UFOs all over the base!

OTHER PROOF

Norton even has some incredible night vision footage of all of the UFOs in their different colors taken on the base. He also has footage of a UFO going from one horizon to another horizon in a mere two seconds. He has video footage of a triangle-shaped UFO that hovers over the base with no sound being emitted from it. He has all of this video footage for any credible UFOlogist to review and which he now ready to share with the right people.

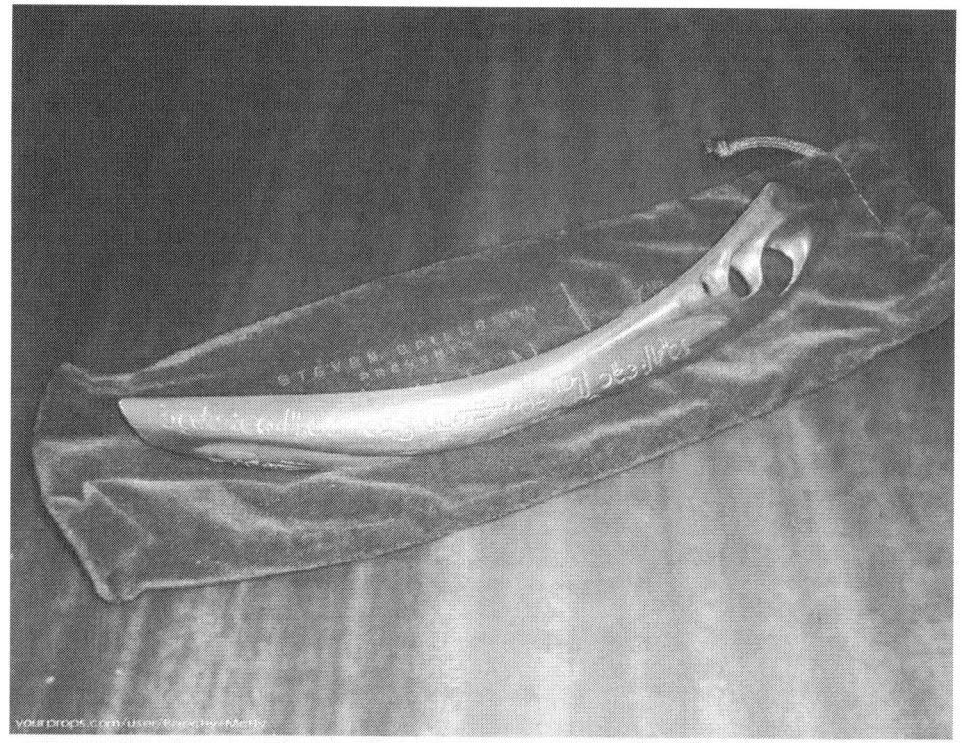

Alleged "proof" of an extraterrestrial encounter. Always keep in mind, if an "alien artifact" seems too good to be true...that may just be the case.

When people of an official rank like senior ranking military personnel, astronauts etc., talk seriously about UFO's and aliens, it's important that the people of this world listen up to what they have got to say!

This is a fascinating story and something which clearly has a lot of question marks against it. We here at U.I.P are going to try to reach out to Norton, to see whether we can be the ones he discloses all of his secrets too!

Something obviously happened that night back in September 1977, and it's very interesting that these multicolored orbs are being seen everywhere nowadays.

CALLER HAS "EXTRATERRESTRIAL" PROOF

The above account is fascinating; however, no real evidence to corroborate it has been made available. That is until researcher Paul Dale Roberts was contacted by a new source who offered "proof" that Command Sergeant Major Norton's story was true.

10/22/2019 - Time: 11:02 AM. Call comes in on the Paranormal Hotline. The caller is identified as Stephanie Fuller. Stephanie discovered an article I wrote about CSM James Norton who witnessed the 1977 skirmish between UFOs and our U.S. military. The skirmish happened at Fort Benning, Georgia.

CSM James Norton claimed to have UFO evidence from that skirmish and to have the evidence buried for safekeeping. One of those objects is the exact same thing that the caller named Stephanie Fuller has.

Stephanie says:

"The item was found in an apartment in Independence, Missouri. My fiancé, Justin Norman, worked maintenance at the apartments. A hoarder had died but it took days for anyone to realize it. Nobody but Justin could stand the smell to clean up the place and the man had nobody to leave his things to, so we were able to keep little things. Anyways, not an extraordinary story but Justin says that he was a WWII veteran and kept to himself."

Stephanie was researching the image of her object on the internet and found it, with my article. CSM James Norton had the same object that Stephanie now has in her possession. Stephanie is willing to talk with any UFOlogist that is willing to analyze her strange object.

UPDATE: 10/29/2019

The object has now been identified; it's not extraterrestrial but is actually a prop from the television show "*Roswell*."

"The mysterious alien artifact that was found at the Roswell UFO crash site at the beginning of the series and was seen extensively throughout the series...Ultimately found to be a somewhat mundane alien recording device. Even so, it was taken away by the aliens, along with hybrid Allie Keys...Chromed resin first generation cast from a screen - used prop."

(www.yourprops.com/Alien-Artifact-replica-movie-prop-Taken-TV-2002-YP58482.html)

Editors Note: This just goes to show you that when it comes to the wild world of UFOs, you can never take "evidence" at face value. Kudos to Paul Dale Roberts for taking the time to properly investigate the so-called "alien artifact" and avoid another piece of spurious evidence from contaminating the already muddy field of UFO reports.

* Paul Dale Roberts is a Fortean investigator who delves into ALL things paranormal – from Mothman, to the Chupacabra, UFOs, Crop Circles, Ghosts, Poltergeists, Demons and more. Roberts is the owner of HPI (Hegelianism Paranormal Intelligence – International).

www.facebook.com/groups/HPIinternational

One controversial UFO artifact was found by Bob White in 1985 near the Colorado-Utah border. Isotope abundance ratio tests show Bob's object has a makeup similar to Martian meteorites.

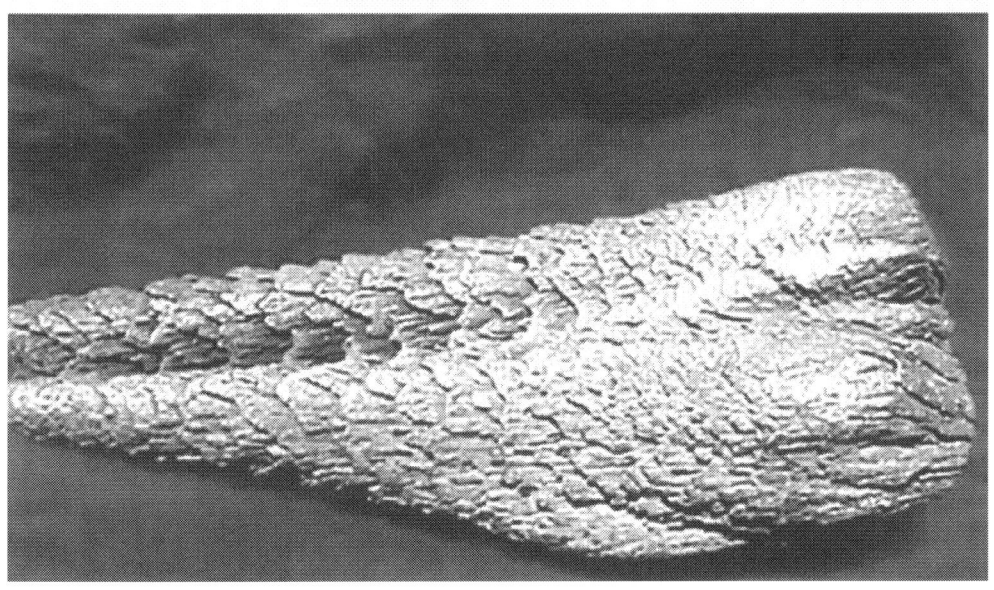

If you have any questions or comments
about this book, please email us at:

commanderx12@hotmail.com

Printed in Great Britain
by Amazon

10491622R00192